Great Forests and Mighty Men

EARLY YEARS IN CANADA'S VAST WOODLANDS

Great Forests and Mighty Men

EARLY YEARS IN CANADA'S VAST WOODLANDS

DAVID LEE

JAMES LORIMER & COMPANY LTD., PUBLISHERS
TORONTO

To Beck, Will, and Joanne.

Copyright © 2007 by David Lee
All rights reserved. No part of this book may be reproduced or transmitted in any form or by any means, electronic or mechanical, including photocopying, or by any information storage or retrieval system, without permission in writing from the publisher.

James Lorimer & Company Ltd. acknowledges the support of the Ontario Arts Council. We acknowledge the support of the Government of Canada through the Book Publishing Industry Development Program (BPIDP) for our publishing activities. We acknowledge the support of the Canada Council for the Arts for our publishing program. We acknowledge the support of the Government of Ontario through the Ontario Media Development Corporation's Ontario Book Initiative.

Library and Archives Canada Cataloguing in Publication

Lee, David
  Great forests and mighty men : early years in Canada's vast woodlands / by David Lee.

Includes bibliographical references and index.

ISBN 978-1-55028-984-8

  1. Logging — Canada, Eastern — History. 2. Lumber trade — Canada, Eastern — History. 3. Lumbermen — Canada, Eastern — Biography. 4. Canada, Eastern — History. I.Title.

SD538.3.C2L397 2007   634.9'80971   C2007-900338-9

Title page: Logs piled at Algonquin Provincial Park's outdoor logging museum
Opposite title page: Logs piled at the riverside before the spring thaw, January 1903.

James Lorimer & Company Ltd., Publishers
317 Adelaide Street West, Suite 1002
Toronto, ON M5V 1P9
www.lorimer.ca

Printed and bound in Canada

Photo Credits

The publisher wishes to thank the curators and interpretive staff at the participating sites for supplying photographs from their own collections and for their co-operation and help during the photo shoots, in particular, Ron Tozer at the Algonquin Logging Museum at Algonquin Provincial Park; Bruce Henbest, Peter Cazaly, and the interpretive staff at Beach's Sawmill, Upper Canada Village; and Barbara van Vierzen at Lang Pioneer Village Museum.

The images on the following pages were photographed by Rob Stimpson of Rob Stimpson Photography at Algonquin Provincial Park: Back CoverL, 3, 5, 15, 20, 28B, 31, 34, 39T, 43, 45, 67T, 72, 84, 89

The images on the following pages were photographed by Jackie MacRae of Behold Photographics at Upper Canada Village Heritage Park: 67, 69, 73, 75, 76, 77

The images on the following pages were photographed by H.A. Eiselt at King's Landing Historical Settlement: 68, 87

Other photographs were supplied by and appear courtesy of the following historic sites and archives.

Library and Archives Canada: Back Cover B (Charles William Jefferys, C-073702), 6 (C-131920), 7 (John Charlton, PA-121799), 8 (William M. Harmer, C-026244), 12 (John Henry de Rinzy, 1993-343-4), 13 (Alexander Henderson, C-008046), 17 (William Harmer, C-025718), 21 (William Harmer, C-025719), 22 (PA-12033), 26T (PA-121808), 28T (William Harmer, C-026520), 29 (Ronny Jaques National Film Board of Canada, WRM 4928), 37 (W.J. Topley, PA-008394), 38 (Philip John Bainbrigge, 1983-47-62), 40 (Charles William Jefferys, 1972-26-792), 44 (PA-032526), 46 (J.E. May Canada Dept. of Mines and Technical Surveys, PA-020330), 47T (Harry Hinchley, PA-123277), 47R (Henry James Warre, 1965-76-64), 48 (Alexander Henderson, PA-028602), 51 (Ellison & Co, C-090135), 55 (W.J. Topley, PA-008440), 58 (William S. Hunter, C-040783), 59 (W.J. Topley, PA-008405), 63 (C-082859), 71 (C-121146), 74 L(John Wycliffe Lowes Forster, 1960-38-2), 80 (W. J. Topley, PA-009298)

McCord Museum: Front Cover (MP-0000.25.901), Back CoverT (MP-0000.25.867), 11 (MP-0000.25.901), 24 (MP-0000-25-867), 23B (William Notman, I-63213), 26B (William Notman, I-63221), 70 (William Notman I-78889), 74R (M930.50.1.334), 91 (MP0000.25.884), 91 (MP-0000.25.884)

Ontario Archives: Back Cover M (John Boyd, C 7-3 2923), 2 (Macnamara Collection, C 120-3-0-0-134), 10 (Reuben R. Sallows fonds, C 223-1), 13 (John Boyd, C 7-3), 18 (Macnamara Collection, C 120-3-0-0-128), 19 (Multicultural History Society of Ontario fonds, F 1405-15-108), 23T (John Boyd, C 7-3), 25 (Multicultural History Society of Ontario fonds, F 1405-15-50), 27 (William Hampden Tenner, C 311-1-0-20-7), 30 (Macnamara Collection, C 120-3-0-0-110), 33 (Macnamara Collection, C 120-0-0-122), 35 (Macnamara Collection, C 120-3-0-0-109), 39R (John Boyd, C 7-3-2923), 41T (Macnamara Collection, C 1903), 41B (Macnamara Collection, C 120-2 s 5061), 50 (Macnamara Collection, C 120-3-0-0-33), 53 (John Boyd, C 7-3 2389), 54 (E.J. Zavitz, RG 1-448-1, 372), 61 (Lt. Pilkington, F 47-11-1-0-114), 62 (John Boyd, C 7-3 18293a), 64 (Macnamara Collection, C-120-3-0-0-148), 78 (Macnamara Collection, C 120-3-0-0-10), 79 (Macnamara Collection, C 120-3-0-0-9)

Algonquin Park Museum Archives: 42 (R. Thomas & J. Wilkinson)

L=left, R=right, T=top, B=bottom, M=middle

Contents

Introduction	7
1 Working in the Bush	15
2 Working on the River	39
3 Sawmilling	67
Conclusion	89
Sites, Museums, and Festivals	93
Selected Bibliography and Further Readings	94
Index	95

Loading timber aboard ships at Quebec City in the nineteenth century.

Introduction

The nineteenth century was the great age of expansion in the Western world. It was the age of the Industrial Revolution and the migration of millions of Europeans to North America. The nineteenth century was also the age of wood, for it is clear that neither of these momentous historical developments would have been possible without this all-purpose building material. Strong, durable, flexible, versatile, plentiful, and cheap, wood was employed to make a panoply of goods. It was used to provide people with small wares like furniture, hand tools, and farm ploughs as well as larger

Squaring a large pine timber near the Jocko River, Ontario.

A stand of pines ready to be felled.

assets such as houses and barns. It was used to construct the wharves and ships that carried people to the New World and build the cities they founded. It was used for the telephone poles, railway ties, and railcars that connected them. It was used to build the factories of the Industrial Revolution as well as some of the machinery within them. It was used to make the millions of boxes, crates, casks, barrels, kegs, and hogsheads needed every year to hold the products of that revolution. It was used to manufacture the wagons and wagon wheels that carried those products to market. Finally, in an age of growing literacy, it furnished the fibre to make paper for the burgeoning book and newspaper trade. The greatest portion of the wood utilized to propel this age of expansion was in the form of ready-to-use, sawn lumber.

Introduction

Fortunately, the nineteenth century brought advances in sawmill technology, making it possible to turn out lumber in large quantities at low prices.

Fortunately for Canadians, the eastern parts of the country had everything needed to meet the constantly growing demand for lumber — prolific stands of timber, useful river systems, and abundant sources of power. The first thing noticed by Europeans arriving in Canada was that it was blessed with some of the richest coniferous forests in the world. Vast tracts of dense spruce and pine (highly prized by builders) dominated the landscape from northwestern Ontario to Cape Breton, along with smaller, scattered stands of valuable oak, cedar, and tamarack. For some people, the forests were so vast they felt they were limitless and could be harvested forever. One veteran lumberman told of the Madawaska and Petawawa valleys in Ontario in the 1870s: "The whole country was covered with an ocean of pine; as far as the eye could reach from any prominent height, nothing was to be seen but a mass of pine tops; in fact one could imagine that it would never be cut or used up." And this was after the area had already been logged for more than forty years.

Although many of the timberlands were in remote areas, they were drained by extensive river systems that lumbermen could use to float their logs to sawmills and then ship their lumber to buyers at home and abroad. Almost every stream was put to use. The most important included the Trent and Severn rivers in Ontario, the Ottawa shared by both Ontario and Quebec, the St. Lawrence and St-Maurice in Quebec, the Saint John and Miramichi in New Brunswick, and the LaHave and Mersey in Nova Scotia. Equally important, the same rivers were able to supply power to run the sawmills. Whether lumbermen relied on power generated by steam or by waterfalls (which were later also used to produce hydroelectricity), they needed dependable supplies of water and they found it in abundance. The potential offered by this wonderful mix of timber, rivers, and power sources was obvious, and Canadians moved quickly to exploit it to supply their local and personal needs for building materials. Soon they were producing more lumber than they needed and began exporting their surpluses to outside markets; in this way, they helped serve the needs of new immigrants as they spread across North America, and meet the demands of the Industrial Revolution as it swept across the Western world. The world was hungry for wood and Canada was ready to supply it.

Supplying wood to the world required a great commitment in manpower, but Canadians were up to the challenge. Every fall thousands of men lined up to spend months living isolated lives in the wilderness, felling trees and hauling timbers and logs. In the spring many of them stayed on to raft or drive the season's forest harvest downriver to the shipping ports and sawmills. In the summer many of the same men were hired on to work in the sawmills, turning out lumber for sale at home and abroad. Whether in the bush, on the river, or in the mills, the work was dirty, arduous, tedious, and dangerous, but there was seldom any shortage of men ready to take it on. These shantymen, loggers, river drivers, and raftsmen became famous for their athleticism, their durability, their bravery, and their swagger. As we shall see, they were, indeed, the legendary mighty men of the great forests.

Masts and Square Timber

Lumber was not the first forest product to be exported from Canada. In the early 1700s, small shipments of masts and square timber were sent to France from

Loggers cross-cutting a large saw-log.

both Acadia and Quebec. After the British conquest in 1763, these products came to be exported in greater quantities. By this time, the best timber had been stripped from the forests of Great Britain, and the kingdom had become dependent on imports from abroad. Great Britain was an island nation totally dependent on its navy for its wealth and security. New England was one of its major sources of pine masts but, with the American Revolution, this supply was cut off. New England pine had a good reputation for strength, flexibility, and size, and the British now had to look elsewhere for a new source of supply. Sometimes they looked for specimens that could produce masts more than 100 feet in height and over 3

Introduction

Oxen hauling a sled along a skid road, about 1910.

feet in diameter at the base. The British were fortunate to find that the valleys of the Saint John (New Brunswick) and (a little later) the St. Lawrence rivers held vast tracts of large pines that were more than able to meet their needs. By the time of the War of 1812 the production of masts had become a major industry in Canada, one that remained profitable until steamships replaced sailing ships later in the century.

Some of the production was used by local Canadian shipbuilders, but most of the masts went to Britain.

Square timber (hewn flat by master axemen) found a ready market in Britain and almost nowhere else. The British used great volumes of square timber in heavy construction projects — for bridges, wharves, mining supports, the factories of the new industrial economy, and for the grand estates of the

Hauling sawlogs along a skid road, 1924.

nobility. Some was used for ship framing and decking (the decks of the giant ocean liners *Lusitania* and *Mauretania*, launched in 1906, are said to have been built of timbers supplied by the J. R. Booth Company of Ottawa). Square timbers were also sawn into boards and planks for smaller applications. A major advantage of hewing timbers square was that they fitted better into ships' holds than unimproved, round logs. For centuries the British had relied on the Baltic countries for most of their square timber. But at the beginning of the nineteenth century, war with France cut off Britain's access to this area. Once again the British turned to their North American colonies for help and once again they found timber that could meet their needs. Pine was the major wood exported, along with lesser volumes of hardwoods such as oak, birch, and ash. Most of the output was shipped to Britain; little square timber was used locally.

Exports of Canadian square timber grew quickly, soon outstripping shipments of masts. The square timber industry played an important role in the Canadian economy through much of the nineteenth century. At first, New Brunswick was the most important producer of timber but Upper and Lower Canada (Ontario and Quebec) surpassed it in the 1830s, much of the output coming from the rich timberlands of the Ottawa Valley that bisected the

Introduction

Loading timber onto a barge, 1910.

two colonies. By the early twentieth century, however, most of the huge trees the industry required had been cut down, and timbering died out.

Sawn Lumber

For centuries, Canadians have sawn logs into lumber, both for domestic use and for export abroad. The first settlers, of course, used the logs they got from clearing their land to build their homes and make their own furnishings. The simplest way for settlers to cut their own lumber was by hand, with a whipsaw. This long saw could be worked from a scaffold or over a pit; two men, one above and the other below, sawed logs into boards. This work was not only unpleasant (the bottom sawyer getting his face full of sawdust), it was also painfully slow and arduous. Soon, however, mechanized lumber mills were built that used water or steam power to move saw blades up and down through the wood. Machinery replaced muscle power and pit saws virtually disappeared in most parts of the country.

The first mechanized sawmill to operate in Canada seems to have been built in Quebec in the 1660s, and was probably powered by water. Another was operating in Acadia (at Port Royal) by the 1690s. In the years thereafter, mechanized sawmills appeared all over the country, most of them at waterfalls, big and small. In the Maritimes, some lumbermen looked beyond water and steam to power their mills. For instance, one of the earliest lumber mills in Nova Scotia (at Dartmouth, 1753) was powered by wind, and a few other wind-powered mills were later erected around the province. Other lumbermen strove to catch the energy of rising and falling ocean tides. Tide-powered sawmills are known to have operated in the nineteenth century at Saint John in New Brunswick, and at Weymouth and Parrsboro in Nova Scotia.

As the Canadian population grew, sawmills proliferated, giving birth to hundreds of lumber towns across the countryside. The census of 1901 reported a total of 1945 sawmills in Ontario, Quebec, New Brunswick, Nova Scotia, and Prince Edward Island. Most of them were small establishments serving the lumber needs of their local communities; most of them sawed wood for only a few months a year, and some operated in combination with flour mills. However, in the nineteenth century a few dozen mills, usually those with the best access to power sources, grew into giant producers. These establishments turned out millions of board feet of lumber every year, largely for sale in Britain and the United States. It was these large-scale, export-driven sawmills that had the greatest impact on the Canadian economy, providing jobs for thousands of workers and fortunes for a few lumbermen. In 1901, sawmilling ranked second only to farming as the most important industry in Canada, in terms of dollars earned.

The lumber industry was constantly on the move as one area replaced another in supremacy. As the forests of each area were stripped of its best trees, lumbermen looked forever westward and northward for new timberlands to harvest. New Brunswick was the leader at first, holding on until about 1860 when Upper and Lower Canada took over. Here, the Ottawa Valley led in lumber output until about 1890. At this point, the untouched forests of northwestern Ontario were opened up for exploitation and that vast region became a new target for lumbermen. Lumbering did carry on in the older areas, but at more modest levels.

This book offers a portrait of logging and lumbering in eastern Canada in the old days — from the beginnings of the forest industry until modern times (about the middle of the twentieth century). It will show how the work involved in producing square timber, masts, and lumber followed a seasonal round, much like work in agriculture. It will show how the industry spread from the Atlantic provinces in the east to Lake of the Woods in northwestern Ontario. It will show how Canadians used both brain and brawn, machine and muscle, to convert their great forests into valuable building materials. Most important, it will portray the lives of the thousands of men whose work built a country.

1 Working in the Bush

When Europeans first arrived in what is now Canada, the kings of France and Great Britain claimed ownership of all the land, as well as all the trees and other resources on it. Anyone intending to take wood out of the forest needed permission from the Crown. By the end of the eighteenth century, provincial governments held responsibility for administering Crown land; only they could sell it or give permission to exploit its resources. In Ontario and Quebec, agricultural settlers bought or were given Crown land to clear for farming, but few timbermen or lumbermen ever purchased land just to cut wood. In these provinces, governmental authority was widely ignored at first, and loggers simply cut wherever they wished. In the 1820s, however, governments asserted their control and installed a system in which licenses were issued for cutting on particular "timber limits" (or "timber berths") that they had set aside and surveyed. A limit-holder would buy a timber limit at so much per square mile and, at the same time, also commit himself to pay "timber duties" (or "stumpage fees") on all the wood he removed; he never owned the limit, he was simply leasing the cutting rights on it. Stumpage fees and the sale of licenses brought substantial revenue to

Broadaxe for hewing timbers square.

A shanty like this one (circa 1875) would have housed all of these men for six months or more.

provincial treasuries. Between 1867 and 1900 the governments of Ontario and Quebec, for example, got twenty to thirty per cent of their finances from the forest industry.

In Ontario and Quebec almost all logging was done on Crown land, but the picture was different in the Maritimes. Here, the right to cut on Crown land was regulated in much the same way as elsewhere, but more than half the logging was done on privately held lands because governments had either sold vast tracts of Crown land outright to lumbermen and other investors, or given them away as subsidies for railway construction. Any lumberman wishing to cut on private timberland that he did not own had to negotiate cutting rights with the landowner. The question of who had cutting rights on which parcel of land was important to everyone. Lumbermen, for their part, were concerned because trespassing on private timberlands or another lumberman's Crown timber limit sometimes led to bloody clashes in the bush. For their part, governments, whether in the Maritimes, Ontario, or Quebec, were concerned

Working in the Bush

Interior of shanty/lumber camp with camboose fireplace (circa 1900).

because poaching timber on Crown land meant less revenue for their treasuries.

Before paying for a cutting license or arranging a deal with a private landowner, many large-scale lumbermen hired "timber cruisers" to scout the woodlands they intended to harvest. These veteran woodsmen evaluated the quality and quantity of the standing timber; some were said to be so expert they could compute, within a thousand board feet, how much lumber a stand of pine could produce. Cruisers also advised on the most convenient routes for log-hauling roads and located the best streams for log driving. With this information, lumbermen were able to decide which limits they wanted to lease as well as estimate their expenses over the next several months' work in the bush and the long-term revenues they could expect to gain.

Bush work usually began in September or October when the lumberman sent a small advance crew — five or six men along with a foreman, a cook, and a few oxen or horses — up to his timber limit; here the men would erect or repair the buildings that would be needed over the coming months. For most of the nineteenth century the workers got by with only three buildings: stables, privies, and living quarters. In Ontario and Quebec the last was known as a "shanty," and in the Maritimes as a "lumber camp." The men who lived in them were

Shantymen, January 1903.

usually called "shantymen" or "lumberers" (the words *logger* and *lumberjack* did not come into common use until the twentieth century).

Shantymen and Lumberers

Throughout eastern Canada, many of the shantymen were small farmers who left the bush as soon as the winter's logging work was done and hurried home for spring planting. A smaller group (mainly men without land) left at the same time to work in the big lumber mills, which were beginning their spring sawing operations. There were some regional differences, however. In the Maritimes, for example, there were many fishermen-lumberers who spent the winter in the bush and the good-weather months at sea. In Ontario and Quebec, there was also a large corps of men who stayed on the lumberman's payroll to work as river-drivers and raftsmen. They considered themselves full-time professionals and had no other occupation; when they finished their work, they spent the rest of the summer months resting and carousing in Ottawa, Montreal, Quebec City, or their home regions. It is difficult to say how many men were working in the shanties at any one time, but an informed estimate made in 1903 put the figure at about 30,000 in Ontario alone.

A great majority of the professionals in the upper

provinces were French Canadians who employers were eager to hire, not only for their skills but also for their loyalty. There was an unspoken feeling that the farther the shantyman was from home, the more likely he was to work from autumn to spring and not leave his employer. As lumbermen exploited new forests in the regions north of Georgian Bay, Lake Superior, and farther west to Lake of the Woods, they filled their camps with francophone shantymen, some coming from as far east as the Gaspé peninsula. Many of the workers were supplied by labour brokers established in Ottawa who specialized in recruiting these professionals. In the 1930s, however, many of the large lumber firms began to send their own agents into Quebec to hire shantymen. At the beginning of the twentieth century, a new source of labour appeared — recently arrived immigrants from Poland, Sweden, and Finland, men who were quick to learn the skills required in bush work. In the Maritimes, three languages might be heard at work in the bush — English, Gaelic, and Acadian French.

For the most part, men from different ethnic backgrounds got along reasonably well in the bush. The only serious discord occurred in the Ottawa Valley in the 1830s, when hundreds of Irish immigrants were left jobless after completing construction of the Rideau Canal. They had no money to buy land for farming, and French Canadians dominated timbering work in the valley. The Irish, who came to be known as "Shiners," were induced by a few unscrupulous timbermen, led by Peter Aylen, to try to drive the French out and take their place. Many Ottawa Valley timbermen at this time paid little heed to the government's licensing system, logging wherever they chose, whether on unlicensed Crown land or, worse, on a timber limit already granted to another operator. If challenged,

Finnish loggers pose for their portrait.

they would simply get the shantymen they employed to impose their will by brute force. The result was violence and blood in the bush. Hoping to improve his own competitive position, Peter Aylen began hiring only Irish workers, even though

Above: Stamping hammer to mark the ownership of sawlogs.
Left: Double-blade axe for felling trees.

they had yet to learn the arts of the timber trade. He fuelled Shiner fervour with lavish offerings of food and whisky and proclaimed himself champion of Irish rights, though this was only a ploy to forward his personal goals of wealth, power, and prestige. Grateful for the jobs and attention, the Irish greenhorns gave Aylen their complete loyalty. In effect, they gave him a private army, which he used to attack and chase away the (mostly French) logging crews of his rivals and destroy their rafts and booms. After some success in the bush and on the river, Aylen turned his attention to Bytown (now Ottawa, then the largest urban centre north of Montreal), where his Shiners harassed citizens on the street and assaulted rival raftsmen as they passed by on the Ottawa River.

Aylen's skulduggery provoked a two-year reign of terror in the Ottawa Valley, but local officials were eventually able to restore law and order. Aylen ceased his scheming and eventually succeeded in building a new reputation for respectability. The Shiners' War (as it came to be called), however, gave Ottawa Valley shantymen a reputation for violence that lasted for decades. Surprisingly, however, peace came to the valley quite quickly, especially after the government installed the sheriffs, courthouses, and jails that had been lacking. Timbermen also learned the advantages of peaceful cooperation over aggressive competition. And further, the Irish managed to learn the skills of the timber trade and, within a few years, timbermen felt comfortable enough to have

Working in the Bush

The men would saw logs into twelve-foot lengths.

Irish and French loggers living in the same shanty and working in the same gang.

Felling and Hauling in the Bush

Throughout the nineteenth century, the main party of shantymen would normally arrive at the timber limit in October, bringing with them a large portion of the supplies needed for the season — logging tools and cooking implements, blankets and clothing, coils of ropes and chains, chests of tea and tobacco, barrels of pork and flour, and loads of hay and oats for the oxen and horses. Most logging operations were carried out in remote, unsettled parts of the country, sometimes hundreds of miles from population centres, so supplying the camps could be difficult. Lumbermen tried to take their supplies as far as possible by steamboat (in New Brunswick, towboats pulled by horses on the riverbank were also widely used), but at some point everything had to be transferred to horse-drawn wagons to reach the camps. Public roads were rudimentary, and lumbermen usually had to build their own "tote" roads and bridges to service their timber limits. Accidents were not uncommon, especially when heavily loaded wagons had to ford streams and struggle up trails hacked out of the bush. Later in the year, when snow covered the ground and the ice on rivers was judged safe, horse-drawn sleighs were put into action, and provisioning the camps became a little

Felling trees with a crosscut saw.

easier. Freight no longer had to be unloaded from steamboats, because sleighs could go the whole distance; they could glide along frozen rivers almost as quickly as steamers in summer and travel much more smoothly on the now snow-covered roads of the back country. Sleighing was efficient but not without risk, however: one year on the Ottawa River, for example, a team of fourteen horses carrying shanty supplies fell through the ice and perished.

As soon as the men were settled in the camp, they set about felling trees. Work in a camp turning out square timber and masts required exceptional

precision and axemanship. A single square timber or mast was an exceedingly valuable product, so great care had to be taken in selecting suitable trees and in felling them. When selecting trees, a timbering gang of five or six men looked for specimens with straight trunks, large girth, few branches, and no outward signs of rot or disease. Many were considered, but few were chosen. The feller was a highly skilled worker who was expected to bring the trees down in the most favourable direction: he tried to drop them where they would not hang up on other trees or sustain damage from rock outcrops, where they would be in the best position for squaring and hauling, and, if possible, on a bed of brush to cushion the fall. For much of the nineteenth century, fellers relied on long-handled, double-bit axes weighing six or seven pounds to do the job. The two-man crosscut saw was not used until the 1870s, when a

Top: Using a two-man saw to fell a tree, Ontario, 1910.
Above: Felling a large pine by axe.

Men rolling logs onto a horse-drawn sled, circa 1925.

number of technical advances were introduced: raking teeth to remove sawdust from the cut, coal oil to wash sticky resin away from the teeth, and wedges to prevent blades from jamming in the cut.

Once the tree was on the ground, the branches and bark were removed and it was "topped off" at a point where the taper became too pronounced to allow squaring. At this point the "liner" stepped in and, using a cord coated with soot or chalk, marked the line along which the timber would be hewn. He was followed by "scorers" who performed the preliminary hewing, chopping the side flat within an inch or two of the line. It was at this point that the hewer, the most skilled member of the gang, took over. With his many years of experience and a ten- to twelve-inch broadaxe sharpened every night to a razor's edge, he carved the giant log flat along two sides. After the gang rolled the timber over, he repeated the process on the remaining flanks. Hewers were the artists of the forest industry, and the best of them were, like master sculptors, able to carve a perfectly smooth surface along the entire length of the timber. In good weather, he and his gang were expected to turn out five to seven timbers a day.

Work in camps that cut sawlogs for lumber mills differed from those that produced square timber and masts (rarely would one camp carry out both activities). A single sawlog had much less value than a squared timber, which was a finished product when it left the bush. Care and skill were not as important in lumber camps, for sawlogs could sustain some damage and still be sawn into lumber. Felling gangs in lumber camps generally numbered only three men. They simply felled the trees, trimmed the branches, removed the tops, and cut the trunks to the desired length (usually twelve to sixteen feet, with a few extra inches called "broomage" to allow for possible damage on the river drive). Each gang was expected to produce more than a hundred sawlogs a day in good weather. Quantity was more important here than quality, strength and stamina more valued than artistry. The skilled work required to produce lumber was performed at the sawmill, not in the bush.

Felling carried on through autumn and winter, but when the snow got deep enough, some men were assigned to begin hauling the winter's forest harvest onto the ice of the nearest stream, down which it would be floated to its destination in the spring. Snow reduced the friction of hauling by covering rocks, roots, and other snags the timber could catch on. In many camps, the workforce was bolstered in December by the arrival of farmers bringing their oxen or horses to haul sleds carrying timbers and sawlogs. This arrangement benefited all parties. For the farmers, it meant cash earnings and free fodder for the winter, when they had little to do around the farm. For the employer, it meant fewer animals of his own to feed the year around. For several decades, oxen did the hauling, but horses

Finnish loggers, with a sleigh full of fire logs, north of Sault Ste. Marie.

Above: Horses hauling a ton or more of timber by sled.
Left: Skidding logs over rough terrain in the bush.

displaced them before the end of the nineteenth century. Oxen could pull heavier loads, eat coarser feed, and withstand rougher treatment, but horses moved faster, ate less, and were more manageable.

Timbermen learned over time that, when it came to hauling square timbers, great care had to be taken. In the early years, these perfectly squared products were simply dragged or skidded by horses along rough trails — soon called "skid roads" — to the water's edge. Despite the snow, these valuable commodities were often damaged along the way, resulting in lower prices when they arrived at the seaport. Eventually, timbermen began equipping their teamsters with double bobsleds, which could deliver the timbers unscathed to the water's edge.

Just getting these massive timbers on and off the sleds was a mighty task. We don't know the size of

Hauling logs using horse-drawn sleighs.

the largest timber ever taken out of the forests of eastern Canada but we do know that one taken out of Peterborough County, Ontario, had a volume of 960 cubic feet — the equivalent of a timber measuring 4 feet square throughout an entire 60-foot length! Manpower and animal-power were employed to shift and lift the timbers. Men needed prying tools, and these evolved over three generations. The first, known as a pike-pole, was no more than a hardwood pole used as a simple lever. In the mid-1800s, timbermen began equipping their workmen with cant-hooks — shorter poles fitted with a metal hook to grip the timbers. A later improvement was the peavey, which was rigged with a hinged, adjustable hook and sharp spike on the end. Horses or oxen were used to hoist the timbers, pulling chains through a system of winches, but this method still required considerable muscle-power from the loggers.

Sawlogs were easier to move over land than square timbers, as they were generally smaller and could withstand rougher treatment. They could be hauled in bulk, so lumbermen developed large, sturdy sleds that could carry twenty-five or thirty logs at a time. This tactic, however, required them to build wider, heavier-duty skid roads. Road-making soon became a major task in the lumber camps. In swampy sections, logs unsuited for lumber were laid

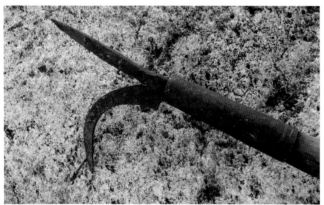

Top: Hauling sawlogs by rail from the bush was expensive. Above: Cant hook for rolling logs in the bush and on the river.

across the road, corduroy fashion, to fill in the hollows. The sled loads, of course, were quite heavy, so hills were a problem. On steep downhills, sand was sprinkled on snow-covered roads to slow the sleds so the horses wouldn't be run over. On uphill grades, steam-powered capstans were employed in later years to help the sleds climb the incline. Sometimes, on level ground, water was sprinkled on the road to form ice and speed their passage.

A few attempts were made in the nineteenth century to mechanize hauling in the bush and do away with horses, but they were not entirely successful. In the 1890s a few lumbermen experimented with a "steam logger," a vehicle that moved on caterpillar

Working in the Bush

Trucks eventually replaced horses for hauling sawlogs, as seen here in the 1940s.

treads and was capable of hauling a chain of log-filled sleds along the skid roads. However, such machines were expensive, heavy, and prone to breakdown, and lumbermen gave up on them. Others invested in short logging railroads with diminutive locomotives to connect their timber limits to mainline railways. With these spur lines, they could move sawlogs from the stump to the sawmill in a few days and in any season. But, although a few were constructed in the Ottawa Valley, they were generally too expensive for most lumbermen. It was the internal combustion engine, not steam power, that replaced horse power in the bush. Trucks began hauling logs out of the bush in the 1920s and ultimately took over all hauling tasks.

Logging was a notoriously wasteful and destructive business. Hauling roads wiped out wide stretches of young growth, the next generation of timber. The trimming of branches and tops left huge piles of slash behind. The square timber industry was a particularly bad offender. Many trees selected as candidates for squaring were discarded after they were brought down when it was discovered they were diseased or rotting at the core. The act of hewing added even more to the piles of waste. Trees are, of course, cylindrical, so the process of

Camboose fireplaces were used for cooking meals and heating the shanty.

hewing them square was especially prodigal: more than a third of the tree trunk, thoroughly sound wood, was cut away and left on the forest floor. In the 1850s, partially squared, "waney" timbers found acceptance in some markets; these sticks were bevelled at the corners, giving them an octagonal profile. But the new approach did not eliminate wastage, it only reduced it. All this waste left in the bush provided perfect fuel for the forest fires that periodically swept over great tracts of timberlands, destroying nearly as many trees as loggers cut down.

Two of the most horrific forest fires in Canadian history took place almost a hundred years apart. The first was in New Brunswick, the great Miramichi fire of 1825 that killed 160 people, almost totally destroyed the towns of Newcastle and Douglastown, and wiped out several thousand square miles of prime forest. Most of the residents along the Miramichi River were connected to the timber trade in one way or another, so the conflagration weakened the local economy for some time thereafter. The worst forest fire ever seen in Ontario burned much of the land between Englehart and Cobalt in 1922, killing 43 people, destroying several towns, and leaving 11,000 people homeless. The fire was blamed on both settlers and loggers. In this case, though, while a great deal of forest was consumed in the fire, some of the best timberlands had already been cut down for lumber.

Algonquin Provincial Park, in its outdoor logging museum, has re-created a camboose.

Life in the Shanty or Camp

Shanties or lumber camps were surprisingly similar from Nova Scotia to Ontario — rough, temporary, multi-purpose structures built of green spruce logs chinked with mud, moss, and bits of wood. The side walls seldom rose higher than six feet, the gable ends only ten feet to the peak. The building was roofed with overlapping "scoops" (logs hollowed out like troughs); by alternating the convex and concave surfaces and resting the edge of one scoop on another, every second scoop formed a channel to carry off rain and melting snow. The floor was formed of roughly hewn timbers. The interior of the structure was dominated by a large open fireplace, in many places called a "camboose," which consisted of a frame of stone or hardwood (six to twelve feet square), holding a foot or so of sand and gravel in which a fire was kept burning day and night. The camboose served for both heating and cooking. The roof above it featured a large opening, topped by a short, wooden chimney that allowed the smoke to escape and, at the same time, let air and light in (most shanties had no windows and only one small door). The walls were lined with double-tiered bunks, in which the men slept with their feet to the fire; thick blankets kept them warm, and spruce boughs served as mattresses. At the end of his bunk, each man would sit and eat on a wooden bench, or "deacon seat," for there were no chairs, and only the cook had a table. The rest of the interior was cluttered with piles of firewood, barrels of water, crates of food supplies, a desk for the foreman, and grindstones for sharpening axes. Shanties ranged in size according to the number of men they were expected to accommodate — from as few as twenty to as many as fifty. They were intended to last only two or three years (when most of the marketable timber within walking distance would have been cut down), and the advance crew sent ahead in autumn could usually build one in a week or so. Construction materials cost little: all the wood was cut on site, and few nails or spikes were used; the only tools needed were axes, saws, and augers. In choosing the shanty site, the foreman had to ensure it was near a reliable source of drinking water, and was on high, well-drained ground where the water could not be contaminated by fecal waste, yet close to the pine groves where the men would be cutting during the winter.

Everyone who lived in the logging shanties was treated equally except for wages. Everyone, including the foreman, ate the same food, drank from the same water dipper, slept on the same kind of bedding, and worked the same hours. But, as in any other workplace, there was a hierarchy of duties to be performed, a hierarchy that could be seen in the payrolls. The range in pay was not unusually wide, and wage differentials do not seem to have caused any discord in the camps. At the bottom of the wage scale were the "general hands," unskilled labourers who did such work as building and maintaining the skid roads. In the middle range were the fellers and teamsters, who spent the winter cutting and hauling the wood. Next came the indispensable cooks and highly skilled hewers. At the top, making about three times as much as a general hand, was the foreman, the personal representative of the lumberman who paid the bills; he had to have the lumberman's full trust, for it was he who made the crucial day-to-day decisions that made the difference between profit and loss. It was the foreman who chose the site for the camp and supervised its construction. It was he who decided what trees should be cut and how the wood should be taken out of the bush. And it was he who was responsible for getting the most out of a labour force whose working and living conditions were far from appealing.

Work in the bush was sometimes dangerous and always arduous. Shantymen spent their days felling and hauling heavy timbers and thus were subject to back injuries and hernias. Still, many continued to work while injured, knowing that not only would they not be paid for time missed but they would also have the costs of room and board deducted from their wages. In addition, their jobs entailed working with sharp, heavy tools, in frigid temperatures, in deep snow, on uneven terrain; this kind of work inevitably led to serious accidents, and in remote timber limits there was no hope of getting quick medical treatment. There were many ways to die doing bush work. Some men were killed by falling trees that landed unpredictably or recoiled unexpectedly at the stump. Some bled to death after gashing themselves with their axes. Still others were crushed by logs that slipped while being loaded onto or unloaded from sleds.

The men worked from dawn to dusk in the bush. For a midday meal, they brought with them a kettle to brew tea and a cooked meal (beans, for example) to eat in the snow. Since it was winter, the working day could be as short as ten hours. This meant, however, that they spent more than half their time in the shanty, with its lack of privacy, crowded conditions, poor hygiene, and other discomforts. Shanties were warm enough but dark and

Eating a noon meal in the bush.

gloomy, as the only light came from the camboose fire and a few candles. Lamps improved conditions when they arrived late in the nineteenth century, but the coal oil that fuelled them also added to the odours emanating from all the unwashed bodies and unlaundered clothing. Lumbermen provided little in the way of washing facilities for their workers, and water was in short supply. Still, the men were not noticeably concerned. One veteran shantyman recalled, "Quite a number would never change their under-clothes or shirts until the clothes wore out, and as to washing their feet, such a thing never came to their minds." Inattention to laundering often resulted in infestations of lice, but the men remained stoical, some insisting that unwashed clothing actually kept them warmer in winter. Despite the poor hygiene, however, shanties were not known as unhealthy places.

Doing hard labour in cold weather requires heavy inputs of calories to fuel the body. The food that shantymen got was plain and monotonous for it was expensive to add piquancy and variety to diets in remote areas. Lumbermen never stinted on quantity, however, and paid their cooks twice what most shantymen got. In the early years, the men lived on little more than salt pork, hardtack (sea biscuits), molasses, dried cod-

Top: The interior of a cookhouse.
Above: Shantymen drank tea by the gallon.

fish, and tea. By the middle of the nineteenth century a few improvements appeared: bread, baked in the sand of the camboose oven, displaced hardtack; beans supplemented the pork ration; and rice and raisin puddings added a dessert to daily meals. Finally, late in the century, newly constructed railways made heavy, bulky, and perishable foods more available in some areas. Beef, for example, could now be brought in on the hoof, adding fresh meat to the camp fare. Other additions included potatoes for stews, peas for soups, butter, sugar, and even canned goods. Shantymen savoured the change in diet, but some employers grumbled that it caused a decline in their health (but probably were just lamenting the extra costs they had to bear). Everything considered, although lumber camp cuisine may have lacked many of the ingredients that nutritionists today think essential to good health, there is no evidence that shantymen suffered from dietary insufficiencies.

The great staple of life in lumber camps was tea. Visitors never failed to comment on the quantity and strength of the tea consumed. One claimed that shantymen insisted on a potable "strong enough … to float an axe in." Another proclaimed that the tea certainly was nowhere like "the effeminate trash" drunk in the cities and noted that most men drank a pound week, and some double that quantity. Shantymen drank tea several times a day, usually with sugar, no milk. This staple may have been the men's main source of vitamin C, saving them from the ravages of scurvy. As well, their midday tea was important in maintaining their body fluids and heat in cold weather. Tea was the only beverage available to the men, after alcohol was

Camboose shanty/lumber camp, November 1900.

banned from the shanties. In the early decades of the nineteenth century, lumbermen provided their men with liberal helpings of rum and whisky, but the trouble these spirits caused in the close quarters of a shanty eventually led to a ban on all alcohol. The prohibition was never enshrined in statute but by mid-century lumbermen all across eastern Canada were voluntarily applying it. It seems, however, that they had to guarantee vast supplies of tea to substitute for rum and whisky. The men worked Monday through Saturday. Saturday night and Sunday was the time to do the things there was no time for during the work week. Few shanties were located close enough to settled areas to allow men to get home for a day, so they had to make the best of their time off. Saturday night was the time to party: the men stayed up much later, playing cards and other games, telling stories, singing, and dancing. Most camps seemed to have a least one fiddler who would provide the music for both clog and square dancing. In some shanties half the men would put on hats and represent female dance partners.

Sundays were quieter, most men getting extra sleep. Some attended to personal chores such as mending or laundering their clothing and cutting new spruce boughs to freshen their bedding. Few men were able to read, but the poor lighting would have made it difficult for those who could. Sunday was also the day the men could replenish their personal supplies from what was called the "van" in Ontario and the "wogan" or "wanigan" in the Maritimes; it was described as an "immense chest, made of the strongest wood, ribbed with iron bands, and secured by a mighty padlock" that only the foreman could open. This miniature store held a selection of goods the men could buy, particularly shirts, trousers, jackets, mitts, socks, boots, moccasins, and soap, as well as lotions (for sore bodies) and "painkillers" (which may have contained a little alcohol or even narcotics). The most common purchase was tobacco for both chewing and smoking (many of the men smoked pipes constantly, with obvious effects on the shanty's air quality). Shantymen often complained that lumbermen took advantage of them as a captive clientele, charging exorbitant prices for goods sold from the van. Employers, in turn, claimed the prices simply reflected the high costs of transportation to remote shanties.

As the nineteenth century progressed, Canadian churches became increasingly concerned about the plight of the men who spent half their year in the bush, isolated from society. Some churchmen were troubled that many of the men engaged in excesses of drinking and fighting on their way home from the bush; some worried that shanty life itself might lead the men to moral degeneracy. The Roman Catholic Church was anxious about shantymen living so many months without benefit of sacraments and religious instruction, and it was the first to take action. In the 1840s it began to sending priests up the Ottawa Valley to preach the Gospel, hear confession (important for men who faced death routinely), and celebrate mass in the shanties. Soon the church established a wide network of missions in many logging areas of Quebec and Ontario. Later in the century itinerant Methodist preachers and other Protestant missionaries began visiting the lumber camps too, though their activities were never as extensive.

At the end of the nineteenth century, Rev. Alfred Fitzpatrick of the Presbyterian Church took another approach to improving life in the shanties (as well as in mining camps and railway work sites). In 1899 he founded the Canadian Reading Camp Association, which twenty years later was renamed and became famous as Frontier College. Fitzpatrick sought to spread the social gospel of his church, especially the belief that education was the key to social reform. The first step in his program was to bring basic literacy to shantymen. He challenged educated young Canadians to join his group and work alongside the men in the bush by day and conduct reading and writing classes at night; his organization would give the volunteers a $10 or $20 monthly bonus beyond the regular shanty pay. In newspaper advertisements he noted that "candidates who speak French are preferred." Fitzpatrick recruited well and gained the support of a good number of employers; the movement spread across northern Ontario, and later into the Maritimes and western provinces.

In a lumber camp where dozens of men lived and worked together for many months, removed from society and its comforts, performing dangerous, hard labour, and spending almost all their non-working hours in the dark, smelly, smoky confines of a log shanty, desertion might be expected. Desertion was a serious threat to employers because it took so long to replace men in remote logging regions, thus reducing production and hurting their profits. In Ontario and Quebec most employers required their shantymen to sign a contract (many of them marked with a simple X) when hired to work in the bush. The standard contract contained a provision in which the worker agreed to "forfeit all wages if I leave the employ before expiration of my agreement, without just cause, or the consent of my employer or foreman." This legal document not only allowed lumbermen to refuse to pay a deserter for any work done before he left but also allowed prosecution for breach of contract. There were, indeed, cases where courts found men guilty of breaking their contracts, giving them the choice of paying a fine or spending time in jail.

Foremen also kept discipline in the shanties. In some areas of Ontario there were reports of foremen chasing down deserters and beating them to keep them on the job. The foreman helped keep peace within the shanties too. The lack of personal space in these quarters sometimes led to fights among the men but open discord was rare. Perhaps again it was fear of the foreman that kept tempers in check. Eventually, the bully approach to discipline ended, however, particularly as the construction of roads and railways made it easier for men to desert and easier to replace those who left. By the 1880s, one veteran shantymen noted that "the old, bullying brute force principle of governing [the shanties] is now about entirely done away with." Brutish foremen, threats of prosecution, and the prospect of forfeited wages undoubtedly discouraged some shantymen from leaving work before the contracts expired, but shantymen willingly returned year after year to toil under conditions they knew would be unpleasant.

Relations between shantymen and their employers, the men who paid their wages, were generally amicable. Many timbermen and lumbermen liked to visit their shanties every year to inspect the operations and mingle with the workers, and the

Shantymen lived together in close quarters for many months.

shantymen usually enjoyed the attention. Only one nineteenth-century example of serious worker-management strife is known. In the 1870s a timberman with limits in the Trent River Valley reneged on the contracts he had accepted and reduced wages in his shanties after the working season had begun. He arrived at the camp with a large force of men to replace anyone who did not like the new terms. After heated words, one shantymen struck the timberman on the neck, and further violence was averted only when the shanty foreman drew his revolver and fired in the air. A third of the original shanty crew agreed to stay on the job at the lower wage level but, in the end, the timberman may have been the greater loser. The replacement workers turned out to be incompetent and the original workers sabotaged production by putting defective timber into the rafts. The timberman suffered a big loss that year and never tried to produce timber again. It was a long time before organized

Madawaska and St. Johns rivers, New Brunswick, between 1836 and 1842.

strikes were seen in the bush as workers' unions found it difficult to coordinate collective action among loggers widely dispersed over remote regions. The first union-organized strike in the bush did not occur until 1926 when 700 loggers walked off the job in the Thunder Bay District of Ontario.

As time passed and lumber camps grew in size (some to more than a hundred men), life in the bush improved. By the beginning of the twentieth century the traditional camboose shanty was extinct in most areas of eastern Canada, its multiple functions replaced by a number of specialized structures. Storehouses were built to shelter food and equipment. The foreman got his own quarters. Workshops were provided for the full-time blacksmiths who had now joined the shanty workforce (to keep the horses shod and chains and sleds in good repair). Cooking, dining, laundering, and sleeping were now all done in separate buildings so the logger was no longer confined to a cramped, gloomy shanty. The new bunkhouses that replaced the shanty had glazed windows to provide better lighting and box stoves for cleaner heating. The buildings combined both traditional and modern materials: logs for the walls but tarpaper and boards for the gable ends and roofs. In 1901 the Ontario legislature passed legislation governing lumber camps, imposing minimum standards for hygiene and bunkhouse ventilation (300 cubic feet of air space per man).

Even without government legislation lumbermen knew they had to improve conditions in their camps. By the beginning of the twentieth century it was becoming more difficult to find men who would tolerate the discomforts and deprivations of traditional shanty life. By this time other jobs were available for men looking for physical work in rural areas, especially jobs in railway construction, mining development, and wheat harvesting. Conditions in logging camps continued to improve throughout the new century and the Canadian tradition of men going into the bush every year to harvest timber carried on.

2 Working on the River

In March, as the days grew longer and the air smelt of spring, shantymen began to dream of getting out of the bush. As temperatures rose, there was little logging work to do, for timber could be hauled only when there was snow on the trails. Attention now turned to the river, where the winter's cut of timber had been piled. Many of the men left the shanties at this time and went directly home, but some stayed on to take the timber and sawlogs downriver to the sawmills and shipping ports. These were the river drivers and raftsmen whose lives were later romanticized in tales that enthralled readers in the big cities. The second season of the forest industry was about to begin. The work would be punishing and perilous, but there were benefits: going downriver brought the men closer to home, and the pay was better than a winter spent felling and hauling in the bush.

One task remained to be done before the ice on the river melted: the men had to stamp every square timber and every sawlog with the limit-holder's identification mark so he could prove ownership farther downriver. Most of the major rivers were worked by several limit-holders, and they would all send their cut downstream at about the same time (when water levels were at their highest). On some

Above: A log-rolling river driver sorts sawlogs in a millpond.
Top: Stamp hammer used to mark wood.

A painting by C.W. Jeffreys interpreting Philemon Wright's first rafting voyage down the Ottawa River in 1806.

rivers they formed cooperatives to share the expense of improving the water flow and agreed on how to sort their timbers and logs. Still, disagreement and theft were inevitable, so ultimately they lobbied the government to help them. In 1870 the federal government passed the *Timber Marking Act*, which brought long-lasting order to the forest industry (and is still in force today, more than a century later). The act compelled everyone who floated timbers or logs on the inland waters of Canada to select a particular mark, which had to be conspicuously placed on each stick of wood and be recorded in an official registry. Over the years, timbermen and lumbermen showed an imaginative sense of artistry in designing the marks: they included a wide range of initials and numerals as well as figures such as hearts, crowns, stars, leaves, and various combinations thereof. Out on the ice, their employees wielded a heavy, embossed hammer to stamp the owners' mark on the wood, though sometimes a scribing knife was used. Much like cattle branding on the western plains, the marking routine allowed each owner to identify his wood in the river.

River Driving

The work of the river drivers (*draveurs* in French) was much the same across eastern Canada. Their assignment was to "drive" their employer's timber or sawlogs downstream past numerous obstacles and rapids to calmer waters, with minimal loss or damage. When the ice began to melt, the first task was to check over the improvements that their employer had made to the drive streams (at least those close to their shanty), for to be successful, timbermen and lumbermen had to ensure that the drive streams they intended to use were "floatable." Removing boulders and straightening channels (by blasting

Top: River drivers sweeping logs downstream, 1903.
Above: River drivers working in the icy waters of the spring runoff.

with gunpowder) or installing chains of timbers known as "sheer booms," kept logs and timbers in the main channels, guiding them away from shoals and dead-end bays. And sometimes dams could help float the season's harvest downriver unimpeded: by raising water levels upstream, dams lifted the wood

Single stick slides were designed to carry timber and logs around rapids to prevent logjams.

above rocks and shallows. Then, when the floodgates were opened, the surging water carried everything on a great tide that usually overcame any remaining obstacles downstream (hence, they were often called "splash dams"). Some drive streams had a series of such dams that were opened one after the other along the way. In the early years, when two or more drive gangs tried to take their wood down a river at the same time, they sometimes came to blows; fights would break out because everyone wanted to get the job done as quickly as possible. Eventually their employers learned that cooperation was a better strategy and they got together to form joint-stock drive companies to regulate traffic on the river and share the costs of improvements. This strategy was carried out on most major logging rivers in eastern Canada, such as the Southwest Miramichi River Log Driving Company in New Brunswick and the Madawaska Improvement Company in Ontario.

Some rivers featured more elaborate (and more expensive) improvements such as slides or flumes, which were heavy-timbered troughs that channelled water, timbers, and sawlogs around the wildest waterfalls and narrowest gorges, preventing nasty logjams. Slides were not common in the Maritimes, but in the upper provinces some were provided as public works by the government, while others were built by lumbermen themselves. In both cases, tolls were charged for every stick of wood that passed through. In profile, these structures were wider at the top than at the bottom and some of them were quite lengthy. A privately owned slide on the Lièvre River in Quebec stretched 635 feet, while another on the Pigeon River in northwestern Ontario bypassed a 110-foot waterfall. A government slide on the Petawawa River in Ontario measured 1346 feet long, while another on the Coulonge River in Quebec stretched 2956 feet (more than half a mile!).

No matter how much attention was paid to improving floatability, timbers and logs inevitably became stranded in dead-end bays or stuck on shoals, rocks, and overhanging branches, so the river driver was expected to shepherd the season's cut all the way to the mouth of the stream. Speed was imperative, for the work had to be done during the short spring runoff, when water levels were at their highest. Without ample April rains, river levels could fall too quickly, leaving the wood marooned until the next year's drive. On some days, if they were lucky, the river drivers perched themselves beside a log slide, pushing the wood, stick by stick, down the flume to safety. On many other days, though, they spent their time walking along the riverbank or wading in the icy water, using cant hooks or peaveys to keep the timber and sawlogs moving. Sometimes they worked while balancing themselves on a log, "birling" or rolling it to remain

Cant hooks and peaveys were used for rolling logs in the bush and on the river.

upright. (River drivers were famous for their log rolling, which allowed them to dance from log to log and even ride logs down rapids, earning them a wide reputation for daring and athleticism.)

Despite their best efforts, however, river drivers occasionally had to cope with logjams, which usually occurred in the narrowest stretches of a river when one large timber turned sidelong to the current, catching all others coming along, forming a dam and blocking the whole channel. The drive foreman (who had usually supervised the winter's work) had to act quickly, for water levels rose behind the barrier and the jam grew into a mountain of tangled sticks. First, he inspected the entanglement to locate

Great Forests and Mighty Men

Logjam on the Montreal River, Ontario.

the "key-log," which, if dislodged, would likely release the jam. Then he called for volunteers who would venture into the quivering mass of wood to remove the offending stick. It was an unwritten rule that no one could be compelled to take the risk, but river drivers valued bravery very highly, and peer approval usually moved some to volunteer.

The foreman had a few options: he could have the men chop the key-logs (for there could be more than one) out with axes or pry them loose with peaveys; on rivers close to settled areas, he might be able to bring in a windlass, tie a rope to the key-logs, and pull them out. Using a windlass was a little less dangerous, but all methods required volunteers to climb into the core of the jam to deal with the critical logs. This act of bravery or foolhardiness entailed undoubted risk, for the men never knew when the mass might give way. When the jam did break up, the result could be explosive. One veteran timberman recalled "what ticklish work it is to be standing upon a jam, or up to your middle in the water alongside of one, when it begins to give way. I have seen a piece of timber perhaps sixty or seventy feet long forced almost clean out of the water." At the first sign of movement, the men had to scramble for their lives, jumping from one lurching timber to another until they reached the safety of shore. That was when drivers had to call on all their poise, agility, speed, and sure-footedness, for if they did not make it to shore, they would be swept away by the surging water or crushed under tons of wood cascading down the channel. The only precaution taken was to fasten ropes around the men's waists, but this tactic was often of little use against such a mighty force. Death was not uncommon. In 1885, seven men were killed trying to break up a logjam on the Mattawa River in Ontario.

Loggers and river drivers left a rich legacy of folk songs but showed no sign of bitterness about their lives. Some of their songs celebrated good times, how their crew was the best and bravest on the river, and their foreman the most knowledgeable. Others told of terrible accidents, immortalizing the brave victims but never blaming the foreman. Indeed, in a few cases the singers held the river driver himself at least partly responsible for his fate. One was Johnny Stiles, who died trying to dislodge a logjam on the Moose River in Ontario. He was the best driver on the river, the song says, "but he always seemed careless and wild." In another song, "The Jam on Gerry's Rocks," many of the men in a river crew were reluctant to try to break a jam on a Sunday. Still, seven volunteered to do the work, and they all died, including the foreman who shared the dangers with them. The song laments their tragic deaths but chastises nobody.

Tackling a logjam was not the only way to die on the river; drowning was always a danger. Few river

Accidents, sometimes fatal, while breaking up logjams were common.

A pointer boat carrying river drivers, 1923.

drivers were able to swim. Some men drowned when their canoes, full of camp supplies (the drive camp had to be moved almost every day to keep up with the logs), capsized as they tried to run rapids. And, despite their athleticism, it was not unusual for drivers to slip off logs or rocks. Even good swimmers would have had difficulty surviving in the icy, fast-moving waters of the spring runoff. (The introduction of caulked boots at the end of the nineteenth century reduced the incidence of drowning; these boots, their soles studded with quarter-inch spikes, gave drivers surer footing when walking on logs. In the Maritimes, however, many drivers continued to wear oil-tanned moosehide moccasins known as "larrigans.") Most deaths on the drive happened in remote areas and, although the foreman allowed the men to recover the bodies, the need to keep the logs moving usually meant the

Working on the River

Above: Pointer boats navigating rapids.
Right: Timber raft on the St. Maurice River, Quebec, 1842.

dead were buried quickly on the riverbank. No inquests were held, nor any formal burial rites observed.

Was anyone responsible for the high rate of injury and death among the men who worked on the river, or was it simply an unavoidable consequence of the work they did? One might speculate that foremen and employers, hoping to save time and money, were cavalier about the risks they

A river drivers' camp, 1863, with canoe; pointers had yet to be invented.

expected the men to take, but there is no way of knowing their minds. Everyone recognized the bravery the men showed while carrying out their tasks in the face of danger. Some newspapers, however, placed part of the responsibility for accidents on the men themselves, suggesting that they had taken unnecessary risks in their work and had occasionally strayed beyond bravery into recklessness. Bravado was certainly not unknown among these workers. Stories are told of river drivers who, for example, showed off their bravery by accepting dares (and sometimes money) to ride logs through white water. Still, while they knew that it could lead to carelessness, bravery remained a deep source of pride for these men.

Work on the drive was not only dangerous, it could be downright miserable. Some drives in Ontario and Quebec were over a hundred miles long and took more than two months to complete, and the men often worked a seven-day week. (The passage of the *Lord's Day Act* in 1907 prohibited labour on Sundays, but the logs had to be kept moving and the act proved to be unenforceable in remote areas.) The drivers' whole day was spent near, on, or in the frigid water. They were, as well, regularly drenched by spring rains and tormented by clouds of blackflies and mosquitoes. At night, they found little solace when they camped. The only shelter was a lean-to or tent pitched beside the stream. If it did not rain, the drivers could maintain a warming fire, but even then their clothes were not always dry by morning. The only comfort they could count on was plentiful nourishment. The cook's role on the drive was even more important than it was in the shanty. He made sure the men got four feedings a day: full meals in the morning and evening, in addition to two lunches to sustain them during the day.

Two ingenious advances in river transport — the "pointer" and the "alligator" — did make life a little easier on the drive (and lowered the lumberman's costs at the same time). In the early years, lumbermen often used birchbark canoes to help on the drive, but they were inherently fragile and ill-suited for work on turbulent, log-filled rivers. Canoes capsized and men drowned. In the 1860s, however, John Cockburn designed a rugged, stable rowboat specifically for the log drive; this new craft, the pointer, quickly replaced the canoe. With their flaring sides, pointers were famously responsive and nimble, easy to move in any direction. With their upswept, pointed bow and stern (hence the name), they could ride over floating logs with ease. With their shallow draft (as little as four inches), it was said they could "float on a heavy dew." And with their sturdy construction, they could be run through rocky rapids or dragged around them. On the spring drive, the drivers used pointers to carry their tents, blankets, cooking equipment, and provisions downriver. The men also used them to reach hard-to-get-at corners of the river to sweep them clean of stranded logs and tow them into the mainstream. A Tom Thomson painting, *The Pointers*, shows three of these rowboats towing a barge carrying men and horses along the Petawawa River in 1916. In the autumn, pointers showed their versatility by carrying supplies up tributary rivers to the shanties; the fifty-foot model, rowed with six oars, could carry two tons of freight.

John Cockburn began his boat-building career in Ottawa but, in 1869, moved to Pembroke, Ontario, where he set up a plant by the shore of the Ottawa River. In some years, he and his family turned out as many as 200 pointers, and this distinctive brick-red rowboat became the aquatic workhorse of the logging industry (and was used by hydroelectric, mining, and pulp and paper companies, as well). The Cockburn family supplied their clients for a hundred years, ceasing production in 1969; of course, many pointers continued in service for years after that. Lumbermen all over Ontario, Quebec, Michigan, and other U.S. states relied on this versatile craft until trucks took over the job of carrying logs to sawmills. The Cockburns never took out a patent on their boat, and other builders turned out knock-offs that competed successfully in the marketplace. No matter who made them, though, pointers were a familiar sight on logging rivers across central Canada. Some may have been used in the Maritimes too, but lumbermen in the lower provinces seem to have preferred square-ended punts they called "flangers."

The alligator arrived a little later in response to a stubborn problem lumbermen faced when trying to move logs along remote rivers and lakes with little or no current. For years they used winches erected on land or anchored barges to pull booms of logs across

Alligator tug winching itself over land on roller logs.

these waters. Steam-powered tugboats would certainly have helped but it was too costly to carry such bulky vessels into the hinterland or to build them there. The alligator was an amphibious craft devised by the firm of West and Peachey of Simcoe, Ontario, and first put on the market in 1889. This steam-powered sidewheeler was capable of transporting itself overland into the back country far from the mainstream of navigation. It came equipped with its own powerful winch and thick steel cable. When the cable was attached to a tree, the alligator could winch itself out of the water and pull itself across swamps and up small hills, portaging from one navigable body of water to another. It wasn't necessary to prepare a fancy trail for the portage, just logs and skids placed across the trail to keep the hull from scraping on rocks or roots. The manufacturers claimed it could easily travel a mile a day in this manner. After arriving in the back country, the alligator served as a tugboat, towing log booms along the still waters of interior lakes. In the autumn it also saw service towing scows carrying three or four tons of freight (and even horses) up inland rivers to supply the shanties for the winter. This versatile new craft brought great savings to lumbermen in Ontario, Quebec, and some nearby states. The costs of shipping a bulky alligator by rail made its usage rarer in the Maritimes, but a few were utilized here too.

The spring drive ended when the men finally got their treasure-load of timber or logs past the river's last stretch of whitewater. The men were paid off and finally released from a winter of drudgery in the bush and a spring of miserable work on the river. Paydays could be turbulent affairs, however, for the men were ready to party and had cash in their pockets. To no one's surprise, these occasions attracted predatory publicans eager to oblige them with whisky (usually of the lowest grade) in makeshift taverns they set up by the riverside. The outcome was often mass drunkenness and fighting. After a few days of carousing, some of the men would be hired to work on the log booms or timber rafts that took the wood farther downriver to the sawmills or Quebec City. The remainder would head for home, by foot, steamboat, or rail. Sometimes the men would stop in towns along the way to pick up more booze, and the result could be more fighting or petty theft (such as chicken coop raids). When they reached larger population centres, they could satisfy other neglected needs — good food, new clothing, baths, and haircuts. After so many months of self-

Ships loading timber, Quebec City, between 1860 and 1870.

imposed exile and deprivation, the money went fast. There are many sad tales of men engaging in wild celebration after leaving the bush and the river, some spending everything they had earned in the last six or eight months. Outsiders decried the excesses, ignoring the valuable contributions river drivers made to the economy.

The end of the river drive brought on the start of the third season of work in the forest industry — a season of rafting or booming. The square timber that had been driven downriver now would be rafted to shipping ports, where it would be loaded onto seagoing vessels and sent across the Atlantic to Britain. The winter's production of sawlogs would be towed in booms to lumber mills where it would be cut up into boards and planks.

Rafting Timber in the Maritime Provinces

In the Maritimes many drive rivers ended conveniently at saltwater ports, where square timber could be easily loaded directly onto ships. It was only on the Saint John, St. Croix, and Miramichi rivers that timber had to be taken farther and rafting was necessary. On these rivers, a number of boom and drive companies rafted the timber to a shipping port. A New Brunswick raft was a collection of smaller units called "joints." To make a joint, four long timbers called "floats" were chosen to form the sides of a frame, which was then reinforced by several more timbers laid transversely across the width; the frame was held together by wooden pins, ropes, or chains. Smaller timbers were added, floating inside the frame, while others were loaded on top (usually hardwood timbers, which couldn't float for long distances). Both man- and horse-power, along with pulleys and chains, were used to lift the massive timbers into and onto the joint. The joints could be of any dimension and were usually rectangular. Upon completion, the joint was "bracketed" together with many other joints (sometimes hundreds) to make up an extended raft ready to undertake its journey. The raft could drift with the current for part of the day, but when the tide turned it had to be anchored by the raftsmen. These men used oars and a sweep rudder to steer it through the difficult stretches of the river. As well, masts were sometimes installed to sail the raft with the wind or steam-powered tugboats were hired to tow it.

The Saint John River carried the heaviest rafting traffic. The greatest difficulty that raftsmen faced on this river was getting through the turbulent waters of the Reversing Falls at the city of Saint John. Here, the rafts were dismantled into smaller, more manageable groups of joints and the raftsmen tried to get them through at slack water (the turn of the low tide). Timing was crucial and sometimes things did not work out as planned — joints were upset in the turbulence and men were drowned. The work demanded both skill and daring and the raftsmen insisted on good pay. For a few years in the 1860s raftsmen at Saint John had an association of some 150 members that bargained collectively with their employers, but it was the only union known to have existed among bush and river workers in the nineteenth century. Figures cited in a labour inquiry thirty years later show that, of all day-wage occupations at Saint John, raftsmen were the third-best paid.

In the Miramichi estuary, rafting was quite different: here, rafts headed to their destinations — the seaports of Newcastle, Douglastown, and Chatham — from two directions. Rafting square timber downriver from the west was easy: the men simply took advantage of the current and the ebbing tide and steered the rafts to the ports. Moving timber from the east, however, was a unique and remarkable feat as the raftsmen had first to cope with changing ocean tides and then make their way upriver against the current. Towing by tugboat was the only solution. The timber on these large rafts had been cut beside — and driven down — rivers that emptied into the salt water of Miramichi Bay or the Gulf of St. Lawrence, rivers such as the Tabusintac, Bartibog, Black, and Bay du Vin. The mouths of these rivers were too shallow to accommodate ocean-going ships, so the timber had to be rafted westward up the Miramichi estuary to be loaded.

Rafting Timber on the Ottawa River

Rafting in New Brunswick was neither a long nor dangerous process compared to rafting in the upper provinces. In the Ottawa Valley square timber was produced as far north as Lake Temiskaming; getting this timber down the Ottawa and St. Lawrence rivers to its destination, the seaport at Quebec City, involved a trip of more than 600 miles, some of it navigating through very rough waters. Between

Working on the River

Tugboats preparing to tow booms of sawlogs.

Lake Temiskaming and Montreal the rafts had to deal with eight major waterfalls (most of them today harnessed to produce hydroelectricity) and long, treacherous stretches of whitewater rapids. After passing Montreal, the rafts carried on down the St. Lawrence, heading for Quebec; on the way they had to cross the shallow waters of Lake St-Pierre, notorious for high winds and high waves.

Rafts on the Ottawa were only slightly different from those in New Brunswick: the smallest units (called "cribs" on the Ottawa) were built more securely and with heavy-duty frames to withstand the buffeting they took as they went downriver; and, while New Brunswick joints could be of any dimension, those on the Ottawa were restricted to a width of twenty-four feet so they could fit into the timber slides built along the river. These slides differed from those used on the river drive, which could handle only one stick of timber at a time; timber slides on the Ottawa River were designed to carry a whole crib around a waterfall. The world's first such slide was conceived and constructed by Ruggles Wright

Raftsmen rowing a small timber raft on the Ottawa River.

in 1829 to bypass Chaudière Falls, which straddle the river between the cities of Hull (now Gatineau) and Ottawa. Before that time, each square timber coming down the Ottawa was either carried by wagon around the major waterfalls or allowed to run through them (at the risk of serious damage). This meant that not only was each timber raft taken apart at every cataract, but so was each crib. With Wright's slide, the rafts still had to be taken apart, but the cribs did not. At times, many rafts had had to queue up for weeks waiting to get past Chaudière Falls; now, on some days, a whole raft could get through.

Wright's marvellous invention brought great savings in time and labour, and other bypasses were soon built on the Ottawa River. In the 1840s the government took over these privately owned slides and went on to rebuild them and construct new ones; by 1874 the government had provided timber slides to bypass eight major waterfalls on the Ottawa. (By this time, it had also built similar slides on the Trent River, farther west in Ontario.) Timbermen paid a fee for each crib that descended a slide.

Government engineers added a few refinements to Wright's original design. For example, at some of the higher waterfalls, rather than building one steeply pitched slide, they built several, more gently inclined models, separated by stretches of level

Raftsmen taking a timber crib down a slide at Ottawa; Parliament buildings in the background.

water; at the Chaudière, the whole work extended for more than a quarter of a mile. Another refinement was to outfit the lowest slide with an "apron," which was hinged at the bottom to rise and sink as water levels fluctuated. If there was no accommodation for low river levels, cribs tended to dive into the still-water at the end of the slide, where they were sometimes torn apart by the impact or by hitting the river bottom. Both these modifications slowed the speed of the cribs on their descent and allowed them to exit more gently.

In the nineteenth century, the timber slide at Chaudière Falls in Ottawa became a tourist attraction of sorts. Future kings Edward VII and George V, as well as governors general, prime ministers and their wives, along with visiting celebrities such as Mark Twain took ceremonial rides down the slide on timber cribs. Of course, extra care was always taken to ensure the crib was tightly secured and only the best raftsmen were chosen to steer them. Nevertheless, running the slide was undoubtedly an exhilarating — perhaps even alarming — experience for many people. In those days, the fastest speed anyone could ever experience was on a railway coach, and the speeds reached on cribs descending the slide came close to those of the fastest trains. More important, the hills of a railway line were gentler than the abrupt drops of a timber slide, and travelling in an enclosed railway coach would have been much more reassuring than riding in the open air, with water spraying up between the timbers.

The principal of Queen's University, George M. Grant, left a stirring account of his experience, around 1880. He begins by telling of the anxiety felt by those on board as the raftsman steered the crib towards the Chaudière slide entrance and the sluice gate was thrown open:

The ladies gather up their garments as the crib, now beginning to feel the current, takes matters into its own hands; with rapidly quickening speed, the unwieldy craft passes under a bridge, and with a groan and a mighty cracking and splashing, plunges nose forward, and tail high in the air, over the first drop. Now she is in the slide proper, and the pace is exhilarating; on, over the smooth timbers she glides swiftly ... Now comes a bigger drop than the last, and the water, as we go over, surges up through our timbers, and a shower of spray falls about us ... Another interval of smooth rush, and again a drop, and yet another. Ahead there is a gleam of tossed and tumbled water which shows the end of the descent; down still we rush, and with one last wild dip, which sends the water spurting up about our feet, we have reached the bottom, cleverly caught on a floating platform of wood, called the "apron" which prevents our plunging into "full fathoms five." We have "run the slides."

For the timberman who owned the raft, choosing a pilot was crucial. Some pilots worked on rafts only, but most were foremen who had directed the timberman's work in the bush over the winter. His responsibilities on the river were even greater than in the shanties, for he was entrusted with getting the payload all the way to Quebec in safety; a single raft could be worth as much as $100,000 (over $1 million today). Without a good pilot the timberman could lose a whole year's investment, and the pilot was paid accordingly — at least double the wages of other raftsmen. The pilot was expected to know all the quirks of navigation on the Ottawa and St. Lawrence rivers, conditions that could change from week to week. He had to make critical decisions along the route, balancing the need to minimize the owner's expense with the safety of the cargo and the welfare of his crew.

The pilot's command could be lengthy, especially if he was taking a raft from Lake Temiskaming. When things went well, it took him two months to get to Quebec. But many things could hamper his progress: storms, contrary winds, accidents, heavy traffic at the slides, and low water levels. Normally, he could expect to be on the river for three to four months. In some years, though, water levels on the Ottawa fell so low by August that rafting was impossible; if a raft had not reached the St. Lawrence by then, it had to be beached somewhere along the river for the winter. The delay would be costly to the timberman, for he would get no money until the timber reached Quebec the following year, but he still had to pay the raftsmen at this point.

A raft put together on Lake Temiskaming might have to be disassembled and reassembled a dozen times along the Ottawa — at the eight government timber slides, as well as at some of the more daunting rapids. If weather and water conditions were good, the pilot might choose not to disassemble the raft entirely but, rather, try to run the rapids in "bands" of two to six cribs. Only single cribs could descend the slides, however, so here the whole raft had to be taken apart. Whether they were being run through a rapids or descending a slide, all cribs had to be manned: at least two raftsmen were required to navigate a single crib, while ten or more might be needed on a band of cribs. The men used oars to negotiate the best approach to a rapids or slide and, once underway, they could also use pike poles for steering. After completing the descent, the men would snub the cribs along the shore and return by land to bring another one down. Getting timber cribs past the numerous obstacles on the Ottawa River was tedious work; at some obstacles, the men could make no more than two trips a day.

Out in the middle of the river, the rafts could drift downstream with the current, but the pilots could employ a number of tactics to speed their progress. On wider stretches of the Ottawa they could hoist sails to catch the wind. On narrower stretches the raftsmen often had to row, pulling on heavy, thirty-foot oars to keep the ungainly craft on course. Sometimes the winds blew too strongly and the crew had to drop anchors to prevent the raft from being swept into a deep bay where it could be trapped for days or weeks. The most useful expedient method of all was to hire a steamboat to tow the timber downriver. Although it would add considerably to their expenses, many timbermen took this approach in later years so they could be more certain their raft would reach Quebec that summer.

As on the river drive, the potential for accidental death and drowning was always present on the rafts. Out on the open river, raftsmen were vulnerable to lightning strikes (two died in an 1896 storm on the Ottawa), and occasionally men were thrown off lurching timber cribs and lost in the fast water. The greatest danger, however, was found at the rapids and waterfalls that impeded navigation. If there was no timber slide to provide a bypass, pilots had no choice but to run the cribs through the whitewater. Sometimes the outcome was fatal, as cribs capsized or were smashed on the rocks. Slides made it possible to bypass the worst waterfalls, but timbermen expected their pilots to save money and time wherever possible; as a result, pilots often decided to decline the bypass and take the cribs down the main channel. Using this option, the pilot could avoid the expense of slideage fees and save time by not having to completely disassemble and reassemble his rafts running the cribs over the waterfall in bands of two or more. But spurning the slide was a dangerous practice, and many lives (and payloads) were lost when pilots miscalculated the risks.

Using a timber slide did not always guarantee safe passage, however. Occasionally a heavy crib hit the water too abruptly at the bottom of the slide, throwing raftsmen into the river. And sometimes even all the raftsmen's skill and accuracy failed to guide the crib to the entrance of the slide: contrary winds and currents could hinder their manoeuvres. If the crib missed the entrance, there was no second chance; it would be swept on by the current and flung over the waterfall. Few raftsmen survived a plunge over the cataracts of the Ottawa River. Ten men were reported lost at the Chaudière in 1835,

Raftsmen being rescued from the brink of Chaudière Falls in 1854.

but there were a few remarkable escapes. In 1854, for instance, a crib carrying eight raftsmen missed the Chaudière slide but stuck miraculously on the brink of the falls. A chasm separated the men from safety on the Hull shore. But a large crowd soon gathered and effected an ingenious rescue. First, the townspeople threw a light cord across the gap, to which was attached a strong rope and, finally, a thick cable. The raftsmen fastened their end of the cable to the crib while the local people built a tripod on shore to secure the other end. Then a ring was slipped onto the cable and sent across the gap with a second rope. Finally, the men tied themselves to the ring and were pulled individually along the cable to safety.

Rafting on the Ottawa was indeed hazardous at times, and occasionally the work could be hectic, but it also included spells of peaceful quietude and opportunities for revelry. If the raft was towed by a steamer, the raftsmen had little to do but relax in the

Working on the River

Mealtime on an Ottawa River raft; a camboose oven is visible in the rear.

sun. If not, they might be called on to man the oars all day long. Still, the raft was not just a workplace, it was also the men's residence for the duration of the journey. Shelter on board was quite primitive. On some rafts each man built his own tiny quarters from wood scraps — not unlike a dog kennel. On other rafts the men were sheltered together in larger makeshift cabins constructed of rough boards. Every raft had one crib set aside to carry a cookery, with a sand oven similar to the camboose found in the shanties. And, just as in the shanties, there was always plenty of food. When a raft left its remote point of origin up the Ottawa, meals usually had no more variety than the basic shanty fare of pork, hardtack, molasses and, of course, tea. But as the raft came upon the more settled areas of the river, the pilot was able to buy fresh eggs, fish, and vegetables to relieve the dietary monotony everyone had endured through a long winter and spring.

Alcohol was, of course, prohibited on the rafts, just as it was in the shanties. However, the opportunity to drink was much more easily found on the

Ottawa and St. Lawrence rivers than in the bush. Most raftsmen had already participated in the river drive and had had their first drinks, but their thirst hadn't yet been fully slaked, and many of the stopping places along the Ottawa were equipped with saloons ready to indulge them. Some of the most nefarious enterprises were found near timber slides and rapids. Here, after running a crib through a slide or rapids, the men would have to pass several grog houses as they walked upriver to board another crib. A traveller passing the Calumet slide only a few years after it opened was appalled to see whisky vendors harassing raftsmen to risk taking a drink as they worked. He said they were "beset at every point by these harpies; and perhaps the sober raftsman, who in the morning ran the rapids with safety, before night loses his caution and his life." Rafting accidents were sometimes attributed to crews impaired by visits to riverside saloons. On other occasions, when bad weather might keep a raft snubbed to shore for several days, the men would venture into nearby towns and villages where they would succumb to the enticements of eager saloon-keepers.

After the raftsmen completed their voyage to Quebec and were paid off, their revelling became even rowdier. In 1901, Ralph Connor published *The Man from Glengarry: A Tale of the Ottawa*, a novel that richly detailed the everyday life of the raftsmen, including their sprees at Quebec. He told of how

> many a poor fellow, in a single wild carouse in Quebec ... would fling into the hands of sharks and harlots and tavern-keepers, with whom the bosses were sometimes in league, the earnings of his long winter's work, and would wake to find himself sick and penniless, far from home and broken in spirit.

Local schemers took advantage of the men as they celebrated and many of the celebrants ended up with no money in their pockets after so many months of hardship, danger, and toil. The author, a Presbyterian clergyman, did not condemn their behaviour but, rather, showed how it was perfectly understandable.

The timber rafts that floated down the Ottawa were a familiar and impressive feature on the local riverscape for a long time. During some years in the nineteenth century, 300 rafts or more (some containing only ten or so cribs and others a hundred or more) would float down the Ottawa. When several arrived at once, they could fill the river from shore to shore. At long range, the raft might give the appearance of a floating island or village, complete with cabins, a cookhouse, and its own population of workers. In rough waters, it could resemble a vast blanket of woven timber, bobbing and heaving on the waves. In quieter waters, it might look like a patchwork quilt stitched together by lace. The picturesque (though not always tidy) Ottawa River raft became the hallmark of the square timber industry throughout eastern Canada.

Rafting Timber on the St. Lawrence River

In the nineteenth century great volumes of square timber were produced along the rivers that flowed into Georgian Bay and lakes Ontario, Erie, and Huron. This western end of the timber industry had its start in the Trent River Valley and a few other tributaries that fed into Lake Ontario. Soon the construction of canals and railways allowed timbering to expand into the hinterlands of Ontario as well as Michigan and a few other U.S. states. The completion of the Welland Canal between lakes Erie and Ontario in 1829 allowed timbermen to take their

Working on the River

Severn River near Georgian Bay.

product by water around the great navigational impediment of Niagara Falls. Some of the square timber produced on the upper lakes was put into booms and towed by steamboats down the lakes and through the canal. Another significant portion was loaded directly onto steamboats or schooners and shipped along the same route. In the 1850s, the construction of railways north of Toronto opened up the rich forest lands of the Georgian Bay watershed, making it easy to ship the timber out by rail. All square timber produced in the Great Lakes watershed was ultimately brought to Garden Island, near Kingston, the beginning of the St. Lawrence River.

The only way to move timber 350 miles down the St. Lawrence to Quebec City, where it could be shipped to Britain, was by rafting, and Garden Island became the starting point for the rafting end of the Great Lakes square timber trade. From here, the Calvin family, who operated grand marshalling yards on the island, dominated rafting down the St. Lawrence for most of the nineteenth century. At first, the Calvins rafted their own square timber downriver, but soon they became a common carrier for other timbermen, shouldering responsibility for taking others' rafts all the way to Quebec for a fee. The Calvins made heavy capital investments in this endeavour: they built and maintained a fleet of steamboats to move timber on the lakes, and their yards at Garden Island included huge steam-powered winches used by their raftsmen to unload timbers from the boats and lift them into and onto the cribs they had prepared. (A typical St. Lawrence River timber crib measured about 42 by 60 feet — much larger than the joints of New Brunswick and the cribs of the Ottawa River.) In the next step, the raftsmen usually fastened four or five cribs together to form a "dram," which could reach 250 or more feet in length. For the long stretches on the journey

A sorting boom on the St. Maurice River, near La Tuque, Quebec, 1923.

to Quebec, they put together a dozen or so drams to form a giant raft, some of which could be over half a mile long.

On the first stage of the journey, the raft was towed seventy miles downriver by steamer to Prescott, where the whitewaters of the St. Lawrence began (before the Seaway was built). From here the current carried the rafts through five fearsome stretches of rapids before reaching calm water at Montreal. There were no steep waterfalls on the upper St. Lawrence, but neither were there timber slides to ease the journey. With an eye on changing river levels, the foreman could choose to run his raft through the whitewater as a unit or disassemble it into its component drams, which would then proceed one by one. At the worst rapids — the Long Sault, Coteau, and Lachine — the foreman would hire extra hands, men with local expertise, to help him get through safely. Once past Montreal, the rafts were usually towed by steamers to the timber coves of Quebec City, where the timber would be loaded onto ships bound for Britain.

In the early days, a rafting trip from Garden Island to Quebec City could take as long as four weeks. But with improved rafting skills and the growing use of towboats, the trip was cut to only about a week by the 1870s. The crews used thirty-foot oars to steer the rafts down the best channels of the river and spent the entire journey on board with their payload. One dram of the raft was set aside to carry a cookhouse equipped with a wood stove, a bunkhouse to shelter the raftsmen, and a separate cabin to house the cook and foreman. The foreman was responsible for the success of the trip and was paid double or triple the wages of the crew he commanded. As on the Ottawa, most of the raftsmen were French-speaking (though there were a few Indians too) hired for the season in Quebec. After delivering the timber to Quebec City, the Calvins would rehire the men and bring them back upriver to Garden Island to crew another raft. St. Lawrence River raftsmen would typically make several trips over the summer.

Booming Sawlogs

In the nineteenth century, lumbermen across eastern Canada worked out elaborate systems of booming, sorting, and towing services to get their sawlogs to their mills, where they would be cut into boards and planks. In most places, lumbermen found booms to be the most efficient way to do the job. New Brunswick was an exception, however, for here some lumbermen brought their logs downriver in the same way as square timber was brought to the seaports: they rafted them in the rigid form of large, rectangular joints. In the other provinces, sawlogs

Working on the River

Tugboat towing booms of sawlogs to a lumber mill.

were corralled loosely inside a loop of boom timbers (larger than the logs they contained) chained end to end; this arrangement, called a "bag boom," was then towed by steam tugs to the sawmills. When the logs reached their destination, they were released into a millpond, where they stayed until they were pulled into the mill to be sawn into lumber.

On the Saint John River, several drive and boom companies worked to bring sawlogs downriver. On the Ottawa, however, one outfit handled the work all the way from the head of Lake Temiskaming to Chaudière Falls, a distance of more than 300 miles. This outfit, the Upper Ottawa Improvement Company, popularly known as the ICO, was founded in 1868 by seven leading lumbermen in the Ottawa Valley. The company grew quickly and by the 1880s was operating a fleet of a dozen steam-powered tugboats. It also employed a large force of "boomsmen" for at least six months of the year. The ICO's payroll grew from nearly four hundred men in the 1880s to over a thousand in 1911.

The ICO charged lumbermen a standard fee each time it handled a log, and a log coming from the upper reaches of the Ottawa could be handled eight or more times in its journey because of all the waterfalls along the way. When a bag boom arrived at a waterfall, the company's men would open it, releasing the sawlogs to run freely through the whitewater; the logs were then collected once again into booms in the still-water below the cataract and

Logs belonging to several lumbermen were sorted so they could be sent to the right sawmill.

towed down the next stretch of the river. The releasing and booming procedures were repeated at each waterfall along the way, the boomsmen living aboard the boats as they went down the river. On some occasions, lumbermen chose to send their sawlogs down one of the government timber slides to avoid the whitewater, though, of course, they had to pay a toll for the service. On the Ottawa, sawlogs were allowed to float freely for awhile and were not sorted until they reached "sorting booms" (chains of timber permanently anchored to rock-filled cribs in the river) farther downstream. There were, indeed, three sorting boom stations: two upriver from Chats Falls and a third just above Chaudière Falls. Here, ICO boomsmen identified and sorted the logs (by the markings stamped on them), using pike poles to manoeuvre the wood in the water. Once the owners were known, the men steered the logs into bag booms according to the sawmill they were headed for. The boomsmen at these stations remained onsite for the duration of the season.

Booming was also done on a larger, international scale both on the Great Lakes and on the Bay of Fundy. In the 1880s some American lumbermen investors purchased timber limits in Ontario but chose to build their sawmills at home. They used powerful steamers to tow vast booms of logs across Georgian Bay and Lake Huron to Michigan, where the wood was sawn into lumber. This practice was harmful to the Canadian economy because the benefits derived from turning a local resource into high-priced lumber were leaving the country. To end these losses, the Ontario government passed legislation in 1898 requiring that all pine cut on Crown land be manufactured into lumber within the province; log booming on the lakes declined but did not completely disappear because enforcement was difficult on these large, remote bodies of water.

In the Maritimes, local investor Hugh Robertson carried out experiments in booming sawlogs by sea in the 1880s. His dream was to tow a huge mass of logs from Joggins, Nova Scotia, to New York City, a distance of 700 miles. Robertson designed a system of steel wires and iron chains to hold the mass together in a way that would resist the force of tides and ocean gales. His first two attempts came apart in storms, but in 1888 he succeeded. The Robertson Raft, as his design was called, was the world's first ocean-going raft, measuring 595 feet long, 55 feet wide, tapering to 10 feet at each end, and containing over 20,000 good-sized logs. Towed by two steamers, it made the voyage to New York in only eleven days, gaining international publicity. Robertson patented the design but never tried again, possibly because the demand for lumber was declining as steel began to replace wood in large construction projects.

The work of a boomsman, especially log sorting, could be dreadfully monotonous. The boom timbers the men operated from were not round but roughly squared and stable, reducing the potential for slipping. Accidents did happen, but working on booms was not inherently dangerous, as it was in logging, river driving, and rafting.

Pride of Vocation

Even though loggers, river drivers, and raftsmen were well aware of the dangerous and disagreeable conditions of the work, thousands of them turned up every September ready to sign on for another year. Some had no other option. The part-timers — who spent the rest of the year farming, fishing, or working in the sawmills — needed winter work to survive. The professionals — those with no other occupation — often come from areas where farming, fishing, and other job prospects were poor. But for others, pay was not the only consideration. For them, work in the shanties and on the rivers may have been an escape from the social restraints of their home communities, a way to enjoy the freedom of the frontier and its all-male culture. (In many ways, the way of life was similar to that of the cowboy on the western plains or the sailor on the high seas.) Other men may have been attracted by the prospect of adventure in the wilderness and a life of physical challenges. In earlier generations, the fur trade had provided similar opportunities in Canada, creating a corps of *voyageurs* who headed into the bush every year. This was not a strong tradition in the Maritimes, but it was in the upper provinces for a time. In 1821, however, the Montreal-based

North West Company was taken over by the Hudson's Bay Company of England, and fewer *voyageurs* were hired. Life in the shanties and on the rivers offered a new outlet for men of independent spirit.

Those who chose this way of life developed a strong pride in their vocation, a pride that helped them endure the nearly intolerable conditions of their work. They could point to their achievements and remind themselves and others that their labours brought wealth to the country. More important, however, they took pride in the noble, manly qualities their vocation demanded: bravery, physical strength, endurance, and cheerfulness in the face of adversity. They seldom showed discontent. The only hint of disaffection was the occasional desertion from the shanties and periodic grumblings about pay. After all, it wasn't manly to complain about working and living conditions. In their own eyes, these qualities made them a special breed of men.

For many years, the general public knew shantymen, river drivers, and raftsmen for little more than their drinking, brawling, and improvidence, but late in the nineteenth century their image changed. Defamation evolved into admiration. The men's contributions to the economy were finally recognized. In addition, their personal qualities were romanticized in literature. Now they were seen as men who lived close to nature, blithely facing danger every day, athletes who could fell a tree in two or three swings of an axe and who could dance on logs as they rode down whitewater rivers. One English visitor was awestruck by the feats he witnessed on a river drive in New Brunswick:

> *I do not know a more exciting scene of its kind than to stand and watch a party of these stalwart woodmen with their long iron-shod poles, jumping from log to log with amazing agility, now balanced on the readily yielding timber, now, with acrobatic dexterity, leaping from one log to another among the noise and clamour of exulting voices, and the fouling and jamming of one log on the other as they crash along the devious windings of the surging torrent.*

In Quebec, writers embellished the deeds of Joe Montferrand (a real-life raftsman who lived from 1802 to 1864) and made him a legend. He was depicted as a giant who embodied the ideals of French-Canadian character: strength, bravery, perseverance, and piety. He was also portrayed as a champion of his people, one who defended his compatriots against the toughest, meanest anglophone bullies on the Ottawa River. In English Canada, Ralph Connor's novel became a bestseller and did much to salvage the raftsmen's reputation, pointing out the dangers of their work and the noble features of their character. Like Joe Montferrand, Connor's hero epitomized the best qualities of his people: he was modest, athletic, brave, and a formidable fighter, though always in a defensive role. Now others agreed — loggers, river drivers, and raftsmen had good reason to take pride in their vocation.

3 Sawmilling

As the square timber industry began to decline in the middle of the nineteenth century, new opportunities appeared that brought even more activity to the forests of eastern Canada. Immigrants from Europe were flooding into the United States, and American forests were being stripped of their trees to meet a surging demand for sawn lumber in an exploding economy. Americans looked northward for help, and a dynamic new industry was born in Canada — large-scale, export-driven sawmilling. Lumbermen in both the Maritimes and the upper provinces had long exported a decent quantity of their goods to Great Britain, and local sawmills had always been able to fill domestic demands for lumber. Now, however, Canadian investors seized the opportunity to expand their operations and jumped into the business of sawing boards and planks for export to the United States.

The American demand for lumber was important but, for a Canadian limit-holder, there were other good reasons to move from timbering to lumbering. By sawing trees into boards and planks, the lumberman wasted far less wood than hewing them square. He could also be far less choosy when cutting trees on his limits; now he could take trees he had bypassed in the past — smaller and poorer quality pines as well as the prolific white spruce were both accepted in sawmills. There were savings to be gained too, for sawlogs did not have to be handled

A huge demand for sawn lumber arose in the mid-nineteenth century.

Sawmill powered by an overshot waterwheel.

lumber per year) were active in Quebec and New Brunswick in the 1830s, but it was the advances in technology in the second half of the nineteenth century that brought the Canadian lumber industry to maturity. The industry was constantly seeking new ways to turn out more lumber, of better quality, more quickly, with fewer workers and less waste. Now investors found the answer in making better use of power sources, mechanizing production and making use of more efficient saws.

Advances in Power Generation

At the beginning of the nineteenth century, except for a few windmills, all lumber mills in Canada were powered by falling water. These mills took water from the river above a waterfall and channelled it down a flume from which it dropped into the buckets of a water wheel. The force of the water filling the buckets turned the wheel, which activated the machinery inside the mill. The water wheel was mounted vertically on the side of the mill and was sometimes enclosed to protect it from ice build-ups (it could not function in freezing temperatures, however). Most mills relied on a single upright saw; moving slowly up and down (it cut only on the down stroke), it could take half an hour to slice one board off a sawlog and the sawyer could eat his lunch while he waited. Despite the slow pace and winter dormancy, these mills had some strong points. They weren't expensive to build or maintain. The mechanics were easy to understand, making repairs manageable. And they could operate on streams with low water flows. There were dozens of these sawmills in every province. But, except for a few large mills in Quebec, their output was quite small and it was aimed only at serving the needs of their local communities.

as carefully as square timber when they were hauled through the bush and driven down rivers. (Square timber continued to be produced, however, but now had to compete with sawlogs for room on many of the rivers of eastern Canada.) By the middle of the nineteenth century the time was ripe for Canadian investors to build large, mechanized sawmills capable of turning out prodigious quantities of lumber for export.

A few sawmills working on a large scale (operations that turned out several million board feet of

Sawmilling

In the 1850s, lumbermen adopted a new means of converting falling water into the power they needed to run their saws. Now the water was channelled down a flume called a "penstock" to the basement of the sawmill, where it hit a horizontally mounted iron turbine; the impact caused the turbine to spin and generate power. Turbines proved to be a major advance in mill technology, allowing production on a grand scale. They generated far more energy than the old water wheels and took up less space. As well, enclosed inside the mill, turbines were less affected by ice and, thus, could run in the milder days of winter. The traditional water wheel continued to power small country sawmills into the twentieth century. But the large-scale, heavily mechanized mills built in Canada after 1850 needed great inputs of power; if they were to rely on falling water for their power, they had to use turbines.

Large-scale lumbermen had another option, however — power generated by steam released from boilers. Steam power in lumbering arrived before turbines. Indeed, as early as 1836, Joseph Cunard was operating a huge steam-powered sawmill at Chatham, New Brunswick, a mill that could turn out 40,000 board feet of lumber per day. In eastern Canada, steam-powered sawmills outnumbered those driven by falling water in every province. Many lumbermen might have preferred to use water power rather than steam in their mills; after all, steam mills were more expensive to build, maintain, and insure (because of the high risk of fires originating in the boilers). But there were just not enough good waterfall sites to go around. Still, steam-powered mills did enjoy a few minor advantages. The most obvious was that they could be located anywhere there was enough water to keep their boilers filled, while water-powered mills needed large and consistent volumes of falling water. And in a few places like Saint John,

Sawlogs were taken directly from millponds such as this one and sawn wet.

New Brunswick, steam-driven mills were able to function nearly the year around.

From a distance, steam-powered sawmills were easily distinguishable from water-powered mills by their smokestacks. Because of the danger of fire, steam was typically generated in a boiler plant built of brick or stone and located away from the mill; pipes took the steam to engines in the basement of the sawmill. In large-capacity mills, the steam plant might hold as

Exterior of a sawmill showing two jack ladders rising from a millpond.

many as a dozen large boilers heated by furnaces using mill scraps as fuel. In contrast, water-driven sawmills were more integrated units, the power usually generated inside the building. In most other ways, however, the two types of sawmills appeared quite similar. Both operated in timber-frame buildings (usually of two storeys) on a thick masonry foundation; they had to be structurally solid to bear the weight of the heavy machinery used in a large-scale operation and withstand the intense vibrations. Inside, on the main floor or deck, the arrangements were also much the same. Both types of sawmills used a complex system of gears, shafts, pulleys, and rubber or leather belts to transmit power from the steam engines or water turbines in the basement to the saws and other machinery on the floor above.

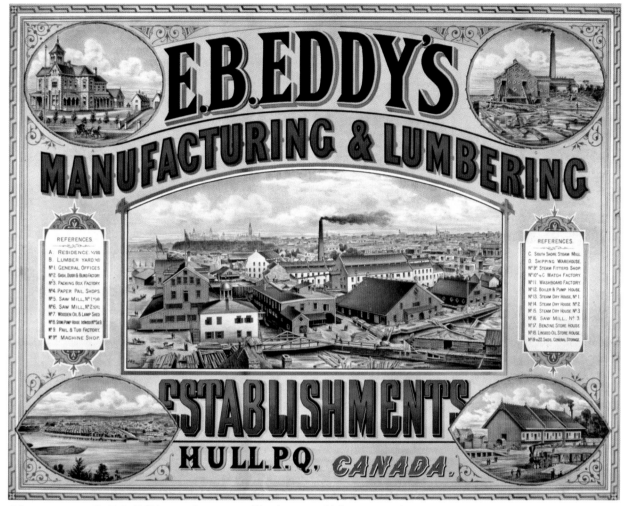

Advertising poster for E. B. Eddy's manufacturing and lumbering establishments, 1884.

The next power option open to lumbermen was electricity, but it did not become available until late in the nineteenth century. The first use of this new power source in a Canadian sawmill was at the E. B. Eddy complex in Hull, Quebec. In June 1881, Eddy switched on forty lights that he had installed inside his sawmill and outside in his lumber yards. The electricity was generated by a turbine driven by water from Chaudière Falls. Some lumbermen had earlier tried using oil-burning lamps to illuminate their mills so they could operate at night during the busy summer season; this was a risky venture, though, because of the possibility of fire. Electrical lighting was much safer, permitting lumbermen to run their saws whenever they pleased. Lighting was one step; powering machinery was another. It took another twenty years

And inside the mill, electrical wiring took up little space compared to the profusion of pulleys and belts they had been using; switching to electricity would provide more efficiency in the layout of machinery and improve working conditions for the mill hands.

Advances in Mechanization

If a lumberman wanted to saw wood efficiently on a grand scale he had to mechanize at least three jobs in the production process. He had to invest in machinery that could pull a continuous run of logs out of the millpond, push them through the saws, then carry the sawn wood around the mill and ultimately outside. As the century progressed, mill engineers designed more and more sophisticated machinery to do the work. Much of the work required heavy-duty machinery but, fortunately for the lumber trade, the new, improved ways of generating water and steam power were able to meet the challenge.

In most mills sawlogs were taken directly from the millpond and sawn wet. In the early years, lumbermen used chains and pulleys to pull the logs up an inclined trough from the water to the saw-deck of the mill. This system was later replaced by a mechanized "jack-ladder"; men standing on booms steered the logs onto this endless chain that snagged the logs and took them up to the deck. Here, the massive sawlog was lifted onto a mechanized carriage and securely clamped, then pushed through the saws. In most mills the outer slabs were removed from each side of the log, then the large remaining "cant" (central core) was turned onto its flat side and sawn into boards. The work of lifting the logs and turning the cants was also mechanized in large-capacity sawmills. Then mechanically rotating cylinders called "live rollers" carried the rough lumber to a "double-edger" saw to have the sides

Some men used oil-burning lamps to illuminate their mills so they could work at night.

before technology progressed enough to allow electricity to be used to run machinery in sawmills and other heavy industries. For lumbermen, the advantages were obvious. With electricity, wiring could carry power over a long distance so a sawmill and its energy source did not have to be in close proximity.

Sawmilling

Lumber was stored in piling yards.

trimmed square; this device had two circular saws, one fixed in place and the other adjustable to the width desired. Finally, more rollers took the boards to a "double-butter" saw where the butts (ends) were cut off to produce boards of the desired length.

Live rollers then cleared the mill deck for more sawing. The fresh lumber, slabs, and refuse wood could choke the mill floor and shut down the production line, so it had to be removed promptly. The lumber was taken outside to the piling yards where it would be stored. In some establishments, the slabs were carried to mini-mills where they were sawn into laths (wooden strips that supported plaster in wall construction). In steam-powered mills, the remaining refuse was conveyed to the boilers where it was burned to generate steam. In water-powered mills, the refuse was burned in incinerators or just dumped into the river where it caused serious pollution.

Advances in Sawing

In the lumber trade, the most striking advances were seen in the development of saws. Innovations such as gang saws, circular saws, and band saws brought wonderful efficiency and greater production capacity to sawmilling in the nineteenth century, making it possible to manufacture huge volumes of lumber at a low cost.

Above: J. R. Booth, whose sawmill at Ottawa was the largest in the world in the 1890s.
Left: A band saw designed for use in a large sawmill.

The first innovation, the gang saw, combined traditional upright saws in a gate or frame, which moved up and down as a whole, cutting a large log into many boards at once. The first known use of a gang saw in North America was in a mill operating near St. Andrews, New Brunswick, in 1803, which worked fifteen saws. Gradually, mechanical engineers were able to design mills with bigger gates and even more saws. By the 1870s, some sawmills were using as many as forty upright blades that could be spaced at one-, two-, or three-inch intervals to cut boards to the thickness desired. Some sawlogs could measure more than four feet in diameter and could take forty saws to cut them. In some cases, though, sawyers (the skilled workmen who controlled the saws) piled four or five smaller logs on the carriage and sent them through the gang together. Gang sawing with uprights was not speedy: it could take up to eight minutes for the saws to eat their way through the length of a sixteen-foot log. It was worth the wait, though, for in the end, the gang saw turned out a wealth of lumber.

The next major step in the evolution of saws was to replace the slow-cutting upright saws with high-velocity circulars. Circular saws had been used for some time for crosscutting, but by the 1850s lumbermen were employing them to rip logs lengthwise

Sawmilling

Working in the sawmill at Upper Canada Village.

into boards. Still, though they could cut wood faster, circulars had a few shortcomings. The lumber they turned out was often rough on the surface, though this was acceptable for lower-grade lumber. And circular saws could cut logs only less than half their diameter. This problem could be remedied by placing one large circular saw above another. In 1902, J. R. Booth used this arrangement to cut the largest sawlog he had ever seen, measuring seven feet across at one end and fifty-one inches at the other. The most serious shortcoming was that the thick blades of circular saws produced "kerfs" (the width of the cut made by the saw) of sometimes more than a quarter of an inch, wasting an enormous amount of wood as sawdust. Despite the problems, however, the speedy new saws were quickly adopted by lumbermen, some of them working them in gangs.

Soon, though, another technological advance changed sawing methods once again: the band saw, a seamless, toothed, ribbon of steel running on two wheels. Small band saws had been used for woodworking since early in the nineteenth century but adapting them to cut huge logs was a big step, accomplished in 1886. Band saws had many

Cutting wood.

advantages. Though they could cut as fast as circular saws, they required less power to operate and the blades had a longer sawing life. They could also turn out smoother, higher-quality lumber. More important, band saws cut a significantly narrower kerf. Some lumbermen claimed that, while a circular could cut nineteen one-inch boards from a particular log, a band saw could turn out twenty-one boards. In a large-scale operation, the savings could be quite meaningful, and lumbermen were quick to install the new saws in their mills. Within a few years, some were using them in gangs. Band saws did not displace circulars, however, as most large-scale sawmills used both types.

Storing and Shipping the Lumber

As technological advances allowed sawmillers to turn out more lumber more quickly, they had to find new, labour-saving ways to carry the increased output to their piling yards, and from the yards to the shipping or railway dock. At first, they used horses to pull wagonloads of lumber in the yards, but many lumbermen went on to improve on this method. As a first step (by the 1870s), they installed rail tracks in their yards and used horse-pulled tramcars to move the lumber. Later, in the twentieth century, some lumbermen brought in steam locomotives to help in their vast yards. For example, the Gillies Brothers' piling yards at Sand Point, Ontario, covered 150 acres and could accommodate 50 million board feet of lumber; the brothers invested in eight miles of track and a thirty-ton locomotive to serve their yards. Tramcars and locomotives, however, only moved lumber around the yards and were no help in stacking the piles or loading and unloading lumber for shipment to market. In this work, all boards and planks had to be handled piece by piece. It was tedious, labour-intensive drudgery that would take a long time to be mechanized.

The immense piling yards seen at every large-scale sawmill held two kinds of lumber. Some of it was wood waiting for sale or shipment to buyers. Most of the stock, though, was lumber piled for seasoning and not yet ready for sale. Seasoning (air drying) reduced the moisture content of the wood, making it less likely to warp or exude gum and, thus, fetched a higher price than green wood. It also made it lighter and therefore cheaper to ship by rail or boat. The air-drying process could take two to six months, depending on the weather, the species, and how green the wood was when sawn; some sawlogs, having spent more than a year in the water (on the river drive and in the millpond), were already partially seasoned on arriving at the saw. In the yards, the lumber was kept off the ground and piled carefully to catch the prevailing winds; room left for

Teamsters hauling a wagonload of sawn lumber to the carpenter shop.

fire-fighting lanes and railway tracks also helped enhance air circulation.

Travellers were always impressed when they came across one of eastern Canada's vast lumber yards, some with stacks twenty feet high. This was particularly so in Ottawa, where several yards were found in the centre of the city, close to the Parliament Buildings. In 1884 one visitor exclaimed,

A great part of the city of Ottawa is a city without residents, a city of lumber. Here are piles of lumber — square, quadruple, diagonal — built

Piling lumber in the McLachlin Brothers' yard, Arnprior, Ontario, between 1900 and 1910.

tier on tier high in the air, lumber for all intents and purposes; acres of inch boards, mountains unending of joists, beams, sheeting, streets of lumber, blocks of lumber, miles on miles of lumber ... Fast as the great mills build the city up, so fast great railway trains and multitudes of immense barges pull it down and carry it away. The air is redolent with the smell of lumber. You breathe pine and resin at every step.

Once the lumber was seasoned, it still had to be sent to market, and some Canadian lumber had to travel great distances to get there. Large shipments went to Great Britain, with smaller quantities also

Loading lumber onto railway cars to go to market, Arnprior, Ontario, 1903.

going to the West Indies and South America. However, in the second half of the nineteenth century, the greatest portion of Canadian lumber exports went to the fast-growing northern United States. Distances became a little shorter in the twentieth century when Canada's own population grew to the point where consumption at home made the domestic market more important than exports. Boards and planks, however, were too bulky and too heavy ever to be shipped very far by road, leaving lumbermen with only two choices of transport — boat or rail. Some depended on both.

Much of Quebec's lumber exports went by rail to New England, though some was still shipped to Britain. Lumbermen in the Ottawa Valley used both means of transport over the years. Initially, they used

Loading lumber from J. R. Booth's sawmill onto barges to go to market.

tugboats to tow three or four barges full of lumber at a time to markets in New England, following the Ottawa and St. Lawrence rivers, the Richelieu Canal, and Lake Champlain. Later, though, some of these lumbermen helped build railways to carry their products to American consumers, totally displacing water transport in the twentieth century. Rail, after all, had the advantage of being able to run in all seasons. Farther west in Ontario, the trip from sawmill to consumer was often taken in two stages — from mill to shipping port, and from shipping port to market port. Hundreds of schooners (and, later, steamboats) sailed out of dozens of small ports on the Great Lakes carrying lumber to the growing cities of the American Midwest. Most ports were served by short, local railways, which brought in lumber from mill towns along the line. In northern Ontario, the construction of the Canadian Pacific Railway late in the nineteenth century opened up vast timberlands beyond Lake Superior; lumbermen operating in these areas found it easy to send their lumber west to build Winnipeg and other new towns sprouting up on the prairies. Railway construction also provided a lively market for railway ties, giving lumbermen an outlet for their poorer-quality sawlogs.

In the Maritimes, Britain remained the most important market for sawn lumber much longer than in the upper provinces; American sales did not take the lead until after 1900. Many lumbermen here were fortunate to be able to build large sawmills at or near good seaports; lumber sawn in Saint John, Chatham, and other coastal towns could be loaded directly onto ocean-going ships and sent quickly off to market (some lumbermen even built their own ships). There were exceptions, however. Lumbermen with mills up the Saint John River used scows to send their sawn products down to the port of Saint John, where they were reloaded into sea-bound vessels. Later in the century, as new railways were built in New Brunswick, sawmills were built far in the interior (closer to the retreating timberlands), allowing lumbermen to send their output overland to American markets. This development brought about a marked decline in lumber manufacturing and shipments from Saint John, which had led the province in output through most of the 1800s. In Nova Scotia, as in New Brunswick, many sawmills were built conveniently close to ocean harbours; but where this was not possible, short railways linked the mills to shipping ports. One fascinating example was the rail connection between the inland mill town of New France and the port of Weymouth. The rails used along the entire eighteen-mile route were made not of steel but of sawn spruce wood! Each rail was twenty to thirty-five feet long and shaped with bevelled edges. The line carried passengers as well as lumber and other freight. Fire destroyed the tracks in 1907, however, and they were not replaced.

Working in the Sawmill

The large-scale sawmills of eastern Canada were major industrial operations, running complex production lines that necessitated an array of specialized jobs. The mill hands were all paid, of course, according to the importance of their work. (Pay varied from year to year and from region to region; it was generally lower in the Maritimes than in Ontario.) At the top of the pay-scale were the sawyers, highly skilled technicians who, by manipulating a set of levers, controlled the whole sawing process; they had to make quick decisions on how best to cut each sawlog as it arrived on the moving carriage. Nearly as important were the boiler engineers (in steam-powered mills) or the machinists (in water-powered operations); their know-how was essential to success in every mechanized sawmill. Another skilled employee was the saw-filer, who kept a broad range of saw blades in good cutting form. The majority of sawmill hands were unskilled workmen such as boomsmen, lumber pilers, and general labourers; they ranked at the bottom of the payroll. Among this group were a surprisingly large number of boys, who performed such tasks as keeping the saw deck clean and piling light-weight laths.

Working in water- and steam-powered sawmills could be bewildering, frightening, and dangerous. High-speed saws and a tangle of live rollers, moving carriages, shafts, pulleys, and belts (some of them two feet wide) offered wide opportunity for serious accidents. Poor working conditions added to the risks — sawdust blowing in the air impaired vision, floors made slippery by water spraying from the saws as they cut through the wet logs made footing uncertain, and openings cut in the floor so mill refuse could be dumped into the river were a threat to safety. As well, numerous distractions — the screeching of the saws as the massive logs were forced into the teeth of the gang-saws and the thunder of the gyrating turbines in the basement — could lessen the workers' attentiveness. The work pressure in the mill could be intense for the men knew that a slowdown anywhere would upset the whole production line.

Young boys working in mills and factories were especially prone to accident, so provincial governments passed laws in the 1880s imposing conditions for their employment. No child could work more than sixty hours a week but, even then, exceptions were allowed in emergencies; some provinces set the minimum age at twelve and others at fourteen. Enforcement was poor, however, especially since the provinces were willing to hire only a few inspectors to keep a check on the workplaces. A few years after the laws were enacted, a royal commission heard testimony from underage boys who had lost limbs in sawmill accidents. One witness who particularly moved the commissioners was an Ottawa Valley boy who had lost both an arm and a leg when he was twelve years old; he testified that the lumber firm (unnamed) had given him $10 for his suffering, but his fellow workers had raised another $25 for him. Child employment declined over the decades, but some underage boys, nevertheless, continued to work in sawmills as mill owners and their foremen ignored the legislation. Indeed, as late as 1913, an eleven-year-old boy was killed at Senator W. C. Edwards' mill in Rockland, Ontario.

Experienced adults were not immune either, and over the years, many were maimed or killed in the sawmills of eastern Canada. In 1893 a newspaper reported a fatality at the Gilmour mill in Hull, in which

William Marquell, an expert band sawyer, had his body literally cut into two pieces. A very large sawlog was being flipped on the carriage when a knot sticking out on it caught the endless chain that drags the logs into the mill from the pond below. The log was given a jerk forward and struck the sawyer's right shoulder. He fell forward on his hands against the moving carriage and the saw cut his arm off below the elbow. It was done in a moment, and the severed arm let his body fall on the moving carriage. He caught the band saw, which ripped him in two before a word could be said or any action taken to prevent the accident.

Employment in sawmills was seasonal everywhere in eastern Canada. The workforce was usually made up of men who had spent the winter in logging shanties and the spring on river drives. No one worked the year round in a sawmill; in most places the season lasted only six or seven months at best — May to October or November. Despite the large number of steam-powered mills, few sawed lumber in the winter, though this was just as much due to difficulties in getting logs to the mill at that time of year as to the cold weather. In the twentieth century, a few lumbermen were able to supply their mills using rail transport and thus saw wood in December. In many sawmills, a full week's employment was rare. Take for example, George Burchill's mill at Chatham, New Brunswick, which was a moderately sized, export-minded outfit, sawing an average of about 4 million board feet a year. In a typical two-week summer period in 1880, sixty-one men were listed on Burchill's payroll; of this number, only forty worked for ten days or more, while some worked for less than one day. The length of a working day declined over the years. In the 1850s, a fifteen-hour day was said to be common at St. Stephen, New Brunswick, but before electric lighting this was likely possible only in the long days of June. By the turn of the century, maximum working hours in most regions had been reduced to ten, six days a week.

Sawmill workers had little success in organizing collective action to bargain with their employers about pay, working conditions, or hours of work. Occasionally, workers managed to shut down lumber production for a day or two to get the employers' attention, but they got no further. Since most mill jobs were unskilled, workers had little clout in pressing their arguments for change. And because the labour force was so mobile, employers could easily bring in replacement workers if they had to. Most important, sawmill owners were not different from other industrialists of the time: they were hostile to any union activity in the workplace. In the early years of large-scale sawmilling, only two actions ever developed into strikes of any duration. In 1891 when the lumbermen of Ottawa-Hull announced a pay cut, 2400 mill hands walked off the job, ending production at eight major sawmills. The mill owners maintained a fairly united front, eventually offering the men a shorter day (ten hours instead of eleven) if they returned to work, but keeping wages at the reduced level. The owners were also able to recruit a few strike-breakers to move lumber out of their yards and convince the authorities to give them police and militia protection. There were no direct negotiations between the two parties, and after a four and a half week standoff, the men quietly drifted back to work, with the lumbermen rescinding the wage cut but maintaining the old working hours.

The second strike, at the James Maclaren and Company sawmill in Buckingham, Quebec, in 1906, ended in three deaths. This tragedy began when the owners locked their employees out of the sawmill after they asked for higher pay and shorter workdays. The mill hands joined together to form a rudimentary union and confronted the company when it tried to move lumber out of the yards. In an ensuing melee, two workers were killed by gunshots fired by security men, employees of a strike-breaking agency. Later, one of the agents also died from injuries incurred in the clash. The local militia was soon called in to restore order in the town. Charges were laid against two Maclaren brothers and some of the agents, but they were ultimately dropped, evidently as a result of political interference. Six workers, however, were convicted of participating in a riot and sentenced to two months in prison. Eventually, after revoking the lockout, the Maclarens drew up a blacklist of over 250 men, who were barred for life from working in the sawmill. With this action, the family succeeded in keeping unions out of their business for another forty years.

Old camp office for the J. R. Booth Lumber Company.

Sawmilling

Timber Barons and Lumber Kings

James Maclaren and his sons were members of an elite group of men (there were no women) whom newspapers, politicians, and the general public referred to as "lumber kings." A similar collection of high achievers in the square timber trade were known as "timber barons." Some — Joseph Cunard on the Miramichi and J. R. Booth on the Ottawa, among others — earned both titles, turning out prodigious amounts of both square timber and sawn lumber year after year. The barons and kings came by their name appropriately, for they reigned over vast empires with large populations. Booth, for instance, at one time owned nearly 3000 square miles in timber limits, an area larger than Prince Edward Island and three times the size of the Duchy of Luxembourg. The power of this aristocracy touched the lives of most people, directly or indirectly, in the areas where they operated. The jobs they offered in the bush, on the drive, and in the sawmills were crucial to local economies where often there was no other significant source of paid employment. The supplies they purchased from local farmers and merchants helped circulate cash in the community. The sawmills they built gave birth to towns and cities. Some of the men added to their power, serving as presidents of banks and railways. Besides the economic clout they wielded, a large number of them added political power by gaining appointment to the Senate and by being elected mayors, members of the provincial assembly, or the House of Commons.

The ruling men of the forest industry came from diverse backgrounds. The largest group were Canadian-born — men such as Jabez Bunting Snowball in New Brunswick, Thomas Gotobed McMullen in Nova Scotia, E. J. Price in Quebec, and J. R. Booth in Ontario. A few were born in England and Ireland but were outnumbered by those of Scottish and American birth. In most cases, those born outside the country became resident in Canada, took out citizenship, and played active roles in the community. The only major group of foreign-based lumbermen were the Americans who invested in sawmills on the upper Great Lakes, many of whom chose to keep their homes below the border.

Most large-scale timber and lumber businesses in eastern Canada were controlled by families or small partnerships. Even in partnerships, however, usually only one man made the important decisions, imprinting his personality on the business, and thus gaining recognition as a baron or king. In many cases, the titles became hereditary, sons succeeding fathers in ruling the empire. In the second and later generations, power was sometimes shared among brothers, but it was a long time before it was shared with stockholders. One of the first big lumber firms to sell shares to the public was Fraser Brothers of New Brunswick in 1919. It was when the business went public that a family lost its rank in the forest aristocracy.

Producing timber and lumber was a risky business, for so many things could go wrong. Too much or too little snow could hamper work in the bush. Low river levels could cause timber and sawlogs to be stranded on the drive or shut down water-powered sawmills farther downstream. High river levels, on the other hand, could break log booms and swamp the mills, also halting the saws. Forest fires could wipe out the timber on the limits the barons and kings had invested in. Sawmill fires (common in buildings constructed of wood and surrounded by piles of lumber and mill scraps) could destroy the costly machinery they had purchased. The timber and lumber they produced were prone to wide and unforeseeable price fluctuations on the international market. The interest charges on the money they borrowed were hard to manage, for it could take from six to twenty-four months (depending on the weather) before their products arrived at market and accounts could be settled. Truly, timber and lumber were classic boom-and-bust industries. Inevitably, in every generation, a number of barons and kings would suffer insolvency and have nothing to pass on to their heirs. Many others, however, posted impressive fortunes. For example, when James Maclaren died in 1892, he left an estate worth $5 million (perhaps $90 million in today's money).

To build their fortunes, the barons and kings relied on integrating their businesses, on personal know-how, good judgment, a little luck, and, indeed, some outright fraud. Most of them ran highly integrated affairs, trying to control as many inputs of their business as possible; some went so far as financing railway construction, becoming involved in banking, and buying or even building ocean-going ships to take their products to market. They also had to know every facet of the business, from the lifespan of the timber limits they were exploiting to the latest developments in sawmill technology. Most of the bankruptcies in their industry came in times of economic depression (especially in the 1840s, 1870s, 1890s, and 1930s); in these times, timbermen and lumbermen had to judge when to be brave and go ahead with production, and when to be cautious and cut back on their activities. Luck often played a role in these decisions too, for prices were notoriously unpredictable.

It is also clear that these men regularly overstepped the lines of the law. In the early years it was common for timbermen and lumbermen to poach whatever wood they wanted from Crown lands without buying a license. In later years, they became experts at finding ways to understate the volume of wood they took from their timber limits, thereby reducing their stumpage fee payments. Both practices resulted in substantial losses to government treasuries. These illegalities were well known in the industry as well as in government circles. In the early 1920s the New Brunswick legislature appointed a one-man royal commission to look into the matter. The commissioner found that the whole system of collecting stumpage fees was corrupt and had been for a long time. He calculated the government's losses in Crown land revenue at several hundred thousand dollars annually ($3 or $4 million today). He claimed that all lumbermen cutting on Crown land were benefiting from "dishonest reductions." In conclusion, he denounced the inaction of the courts, attributing it to the inordinate influence lumbermen held over politicians, some of whom were lumbermen themselves. The corruption found in New Brunswick was common in every province, but no timber barons or lumber kings were ever convicted of fraud. None seems to have shown any embarrassment in breaking the law, simply viewing their illegalities as part of doing business. As the Royal Commissioner said, their behaviour "is not considered any great wrong in the part of the country where they operate."

When sawmills were built in remote areas, it was often necessary for lumbermen to build a whole town to provide for the workers and their families. Some viewed the challenge as an opportunity to control all local commerce. A few operated company stores in their mill towns, but W. C. Edwards, of Rockland, Ontario, went further; he instituted a system of "truck pay," whereby workers were paid in "scrip" (similar to Canadian Tire money), which could be redeemed only at his company store. The

Sawmilling

The C.B. Ross Sash and Soor Factory: the second floor entry was used for moving materials in and out.

St. Anthony Lumber Co. went even further, seeing the challenge as a way to assure itself of a compliant, reliable workforce. The town it built at Whitney, Ontario, in 1896 was intended to be a fully controlled community from the outset. The company owned all property in the town, and non-workers were denied residence. The sale of alcohol was prohibited and drunkenness brought instant dismissal. Single men were accommodated in company boarding houses; married men were provided with cottages and large garden plots that they were strongly encouraged to cultivate.

There were other lumbermen, however, who showed genuine benevolent concern for their workers. One was Arthur M. Dodge, who built a large sawmill and town at Waubaushene, Ontario (on Georgian Bay). He too prohibited whisky sales but allowed private retailers to operate in the town. He may not have set out to build a model town, but he did not stint on providing amenities. The town, which had a population of over 1800 by 1890, featured one and a half miles of wooden sidewalks, a community hall, public bathhouse, churches, and schools (even a high school). He also provided attractive houses, with yards big enough for gardens, for married workers where they could live rent free.

Dodge himself lived in New York City but built a lavish estate on the edge of Waubaushene. His extended family and their servants would spend two or three months there every summer, though they never mingled with the townspeople.

Some barons and kings lived quiet, abstemious lives, with little suggestion of ostentation, apart from their residence. Others often acted in a flagrantly baronial manner. One example was Joseph Cunard, younger brother of Sir Samuel Cunard, founder of the international shipping line. Joseph Cunard was one of the leading timbermen and lumbermen of northeastern New Brunswick in the 1830s and 1840s, sending great volumes of wood to Britain every year in ships he built in his own yards. Cunard was a man who loved attention. Living at Chatham on the Miramichi River, he maintained a majestic, luxuriously furnished home complete with peacocks parading around the grounds. When he attended church, he rode in a coach with the coachman and footmen in full livery. He expected deference and reverence from the townspeople, most of whom were directly or indirectly dependent on him for their living; appropriately, the workers would oblige by saluting him with cannon shots, bonfires, and church bells when he returned from trips abroad. In 1847, however, recklessness and over-expansion pushed Cunard into bankruptcy and he had to close his sawmills and shipyards. Hundreds of people were thrown out of work and the public mood quickly turned hostile. An angry mob confronted the lumber king on the street, threatening him with death, but he armed himself with pistols and stood them off. Cunard eventually moved to England, where he set up a modest ship commission business. Meanwhile, however, his bankruptcy plunged the whole economy of northeastern New Brunswick into a depression, and people left the area by the shipload.

Cunard's story is exceptional, but even the failure of his business shows the power that a single lumber king could command.

For over one hundred years, lumber kings were among the leading industrialists in Canada, known far beyond their local communities. These men were the public face of logging and lumbering. They also symbolized the old ways of doing things. By the middle of the twentieth century, however, most of them were gone, replaced by faceless, publicly owned corporations. At the same time, the old ways of doing things in the forests and the sawmills had also disappeared. The industry had entered the modern age.

Conclusion

During its peak years, the forest industry enjoyed a good public image across the country. But towards the end of the nineteenth century, as many of eastern Canada's richest timberlands were logged out, some Canadians began to question the overall value of the industry. Some, for example, criticized the industry's wasteful logging practices. Others asked why, although they had removed tens of millions of trees from Crown land, the lumber kings had not planted any in

In the second half of the twentieth century, mechanical skidders made bush work more efficient.

return. Others wondered if the industry, rather than stimulating growth in agriculture (as had originally been thought), might actually have harmed it; they felt that too many farmers spent too many months in logging shanties to the neglect of their fields and livestock. Still others pointed out the serious pollution problems caused by lumbermen dumping sawdust, bark, and other refuse from their mills into the nation's rivers, particularly the Ottawa and Saint John.

There is no question, however, that the downsides of timbering and lumbering were far outweighed by the benefits they brought to the country. Indeed, Canada could not have been settled by Europeans without making use of its rich endowment of timber. In the nineteenth century, for example, tens of thousands of immigrants would not have been able get to Canada if it had not been for cheap ocean transport linked to the timber trade: ships built in Canada to carry timber to Britain carried settlers here when they returned to pick up more wood. Once here, everyone found that the forests could supply cheap materials to build homes, barns, bridges, and businesses. The sawmills that produced the building materials gave birth to countless towns and villages. Harvesting the forests provided thousands of jobs and fostered commerce across the country. Stumpage fees helped provide government treasuries with the funds needed for the public works and services on which people depended. The timber and lumber sold abroad helped keep the country's import-export accounts in balance. All in all, without the rich timberlands and the will to harvest them, Canada would have been a much lesser country.

Modern Times

The old ways of logging and lumbering faded away in the second half of the twentieth century as technological advances completely transformed the industry. In the bush, for example, chainsaws displaced the crosscut saw in the 1950s. Soon after, heavy-duty mechanized harvesters and skidders arrived, making it possible for bush work to be carried on with greater efficiency and in all seasons. The new technology has made the work simpler and safer, and government-mandated protective gear such as helmets, goggles, ear mufflers, and steel-toed boots have reduced accidents significantly. The new ways of doing things have also made life easier for loggers. These days, good roads make it possible for the men (and sometimes now, women too) in many areas to live at home and commute to their jobs in the bush by car or truck. Moreover, since logging is no longer seasonal, they can often gain employment the year around. Still, despite the modernization of the lumber industry, sawlogs are no longer the loggers' primary target in the bush; today, more timber is harvested for the pulp and paper industry (which began in the 1890s) than for sawmills. Lumber is no longer as important in the construction industry as it was in the old days; reinforced concrete, steel beams, sheet metal, and even plastics have come into much more common use.

Advances in technology have brought about great changes in sawmilling too. By the middle of the twentieth century, for example, electricity, gasoline, and diesel had become the favoured sources of energy for sawmills. This development brought about revolutionary change in sawmilling, for now

Conclusion

Tractor pulling logs near Iroquois Falls, Ontario, about 1925.

mills no longer had to be located near water. Rather than taking the trees to the saw, the saw could be taken to the trees, even deep in the bush if desired. Lumbermen built new sawmills closer to the timberlands they were harvesting and, as a result, most of the old, established mills closed down. Sawlogs still often had to be hauled to the mill, but the distances were now much shorter and trucks did the work. Trucking, of course, meant the end of that great Canadian tradition, the sawlog river drive. (Pulpwood continued to be driven on some rivers for a longer period, however.) In any case, river driving had become increasingly difficult over the years as the construction of large hydroelectric dams slowed the movement of the logs. Today, these dams have submerged the rapids on which the river drivers and raftsmen had made their fame, creating smooth waters for pleasure boating. Recreational rafting is now pursued on some of the remaining whitewater stretches, though with much more safety than the legendary raftsmen enjoyed.

The new sawmills were usually built of brick or reinforced concrete and roofed with metal sheeting, all to provide greater protection from the fires that had plagued the highly flammable wooden mills of the past. Outside the sawmill, piling yards of considerable size were still common, but mechanized forklifts now took the laborious task of moving, stacking, and loading the mill's output. The move of sawmills closer to the timberlands of course meant that the lumber had to travel a longer distance to reach its markets; railways continued to provide some of the transport as in the past, but the construction of modern highways allowed trucks to do most of this work.

Technological advances have not been the only force in transforming the forest industry. Early in the twentieth century governments passed legislation to clean up rivers by prohibiting the dumping of refuse from sawmills. In the second half of the century, governments turned their attention to the forests, making logging companies follow a "sustained yield" system on Crown lands, hoping to prevent over-cutting and ensure that timber harvesting could be carried on long into the future. By this system, scientists try to compute the amount of standing timber in the forest and forecast its future growth. They then calculate how much timber can be cut without exceeding the amount that will be replaced by the new growth. Governments have also insisted that silvicultural systems be practised on Crown lands in order to boost the quantity and quality of the forest stock. Following this practice, thousands of acres of cutover lands are reforested every year with the species of seedlings that lumber and pulp companies prefer for future harvests. Sometimes, stands of unsuitable species are even removed from areas to be reforested. All in all, the rustic days of logging have given way to modern techniques of scientific forest management.

The old days of logging and lumbering are undoubtedly gone forever, and few artifacts of this colourful past survive outside of museums. The shanties, the sleds, the timber slides, the pointers, the alligators, and most of the old sawmills were built of wood and have nearly all rotted away or been destroyed by fire. One important legacy of the old days has survived, however. The memory of the hardy men who provided the muscle power of the industry lives on in the imagination of Canadians. The swashbuckling feller who brought down the lofty pines. The skilled axeman who could hew timbers square with the precision and art of a sculptor. The brave (or foolhardy) river driver who risked death as he crept into the heart of a logjam to dislodge the key log. The macho, devil-may-care logger who could dance across rolling logs as they hurtled down cascading rivers. The colourfully attired raftsman who caroused his way down the Ottawa and St. Lawrence rivers all the way to Quebec City on a timber raft. Today, Canadians celebrate (and try to emulate) the exploits of these men in lumberjack competitions and logging festivals held across the country. The old days will not be forgotten.

Sites, Museums, and Festivals

Algonquin Logging Museum
inside east gate of Algonquin Provincial Park
P.O. Box 219
Whitney, ON K0J 2M0
705-633-5572
www.algonquinpark.on.ca
The museum portrays the story of logging from the early square timber days to the end of the great spring river drives. Video presentation included. Outside, visitors can inspect a recreated camboose shanty and an amphibious alligator tug.

Blind River Timber Village Museum
P.O. Box 628
180 Leacock Street
Blind River, ON P0R 1B0
705-356-7544
www.timbervillage.ca
The museum houses a large collection of logging and lumbering memorabilia. Visitors can enjoy both indoor and outdoor exhibits.

Central New Brunswick Woodmen's Museum
6342 Highway 8
Boiestown, NB E6A 1Z5
506-369-7214
www.woodsmenmuseum.com
This fifteen-acre park includes models of buildings associated with logging and lumbering, as well as exhibits of woodsmen's tools.

Historic Sherbrooke Village
P.O. Box 295
Sherbrooke, NS B0J 3C0
1-888-743-7845
www.museum.gov.ns.ca
The village features McDonald Brothers' Sawmill, a fully operational, reconstructed sawmill powered by a vertical water wheel. It also has a reconstructed lumber camp.

Kapuskasing Lumberjack Festival
25 Millview Drive
Kapuskasing, ON P5N 2X6
1-800-463-6432
www.kaplumberjack.com
Held every July, this festival celebrates the forest industry with competitions in hand-sawing, chain-sawing, axe-chopping, axe-throwing, pole-climbing, and log-rolling for both amateurs and professionals.

Kings Landing Historical Settlement
20 Kings Landing Road
Kings Landing, NB E6K 3W3
506-363-4999
www.kingslanding.nb.ca
This outdoor museum has an operational reconstructed sawmill of 1830 powered by a 19-foot waterwheel. Examples of logging tools can be found in the sheds of many of the restored houses. Kings Landing also has a working sawmill.

Marten River Provincial Park Visitor Centre
Marten River, ON P0H 1T0
705-892-2200
www.ontarioparks.com/mart
The centre features a reconstructed nineteenth-century logging camp with its log buildings and numerous artifacts.

Musée du bûcheron
780, 5e Avenue
Grandes-Piles, QC G0X 1H0
819-538-7895
www.museedubucheron.site.voila
The museum offers an interactive guided tour on the history of logging and river driving as well as twenty thematic camps on related topics such as a sawmill.

Upper Canada Village
13740 County Road 2
Morrisburg, ON K0C 1X0
1-800-437-2233
www.parks.on.ca/village
The village includes Beach's Mill, a fully operational, water-powered sawmill with a single vertical saw — a typical, small, country sawmill of the nineteenth century.

Selected Bibliography and Further Readings

Angus, James T. *A Deo Victoria: The Story of the Georgian Bay Lumber Company, 1871–1942* (Thunder Bay: Severn, 1990).

Bertrand, J. P. *Timber Wolves: Greed and Corruption in Northwestern Ontario's Timber Industry, 1875–1960* (Thunder Bay: Thunder Bay Historical Society, 1997).

Calvin, D. D. *A Saga of the St. Lawrence: Timber & Shipping through Three Generations* (Toronto: Ryerson, 1945).

Connor, Ralph. *The Man From Glengarry* (Toronto: Westminster, 1901).

Fraser, Joshua. *Shanty, Forest and River Life in the Backwoods of Canada* (Montreal: Lovell, 1883).

Grant, George M. *Picturesque Canada* (Toronto: Belden, 1882).

Hughson, John W., and Courtney C.J. Bond. *Hurling Down the Pine* (Old Chelsea, QC: Historical Society of the Gatineau, 1964).

Johnson, Ralph. *Forests of Nova Scotia: A History* (Halifax: Four East, 1986).

Lee, David. *Lumber Kings and Shantymen: Logging, Lumber and Timber in the Ottawa Valley* (Toronto: Lorimer, 2006).

Lower, A. R. M. *Great Britain's Woodyard: British North America and the Timber Trade, 1763–1867* (Montreal: McGill-Queen's University Press, 1973).

_____. *The North American Assault on the Canadian Forest* (Toronto: Ryerson, 1938).

_____. *Settlement and the Forest Frontiers of Eastern Canada* (Toronto: Macmillan, 1936).

MacKay, Donald. *The Lumberjacks* (Toronto: McGraw-Hill Ryerson, 1978).

Myles, Roderick C. *The Shanty Boy* (North Bay: Myles, 1982).

Parker, Mike. *Woodchips & Beans: Life in the Early Lumber Woods of Nova Scotia* (Halifax: Nimbus, 1992).

Priamo, Carol. *Mills of Canada* (Toronto: McGraw-Hill Ryerson, 1976).

Radforth, Ian. *Bushworkers and Bosses: Logging in Northern Ontario, 1900–1980* (Toronto: University of Toronto Press, 1987).

Robertson, Barbara R. *Sawpower: Making Lumber in the Sawmills of Nova Scotia* (Halifax: Nimbus, 1986).

Thompson, George. *Up to Date, or the Life of a Lumberman* (Peterborough: Times, 1895).

Wynn, Graeme. *Timber Colony: A Historical Geography of Early Nineteenth Century New Brunswick* (Toronto: University of Toronto Press, 1981).United States, 14, 67; as lumber market, 79, 81

Index

Acadia, 10, 13
accidents, 21, 22, 32, 45-46, 47-48, 56, 57-58, 60, 65; reduction of, 90; in sawmills, 82
agents, 19
alcohol, 34-35, 50, 59-60, 87
alligator tug, 49-50
American Midwest, 81
American Revolution, 10
Americans, 85; and demand for lumber, 67; and removing of timber, 65
Arnprior, Ontario, 30, 35, 78
ash trees, 12
axemen, 23, 92
axes, double-bit, 20, 23
Aylen, Peter, 19-20

bag booms, 63, 65; and waterfalls, 63-64
band saws, 74, 75-76
bands (of cribs), 57
bankruptcies, 86; effect on economy, 88
barges, 49
birch, 12
birling, 43
blacksmiths, 38
boat transport of lumber, 79-81
boiler engineers (in steam-powered mills), 81
boom and drive companies, 52
booms (logs), 50-51, 53, 62-63; and Welland Canal, 61; ocean-going, 65
boomsmen, 63-64, 65, 82
Booth, J.R., 74, 75, 80, 85
bracketing, 52
bridges, 21
Britain and the British, 10, 11, 15, 51, 61, 62, 66, 67, 85, 90; as market for sawn lumber, 78, 81; as market for square timber and lumber, 11-12, 14
broadaxes, 24
broomage, 25
bunkhouses, 38; conditions in, 38
Burchill, George, 83
bush, 21-38; life in, 31-38; workday in, 32, 33, 35
Bytown, 20. *See* Ottawa

C.B. Ross Sash and Door Company, 87
Calumet slide, 60
Calvin family, 61, 62
caboose shanties, 35, 38. *See also* shanties
cabooses, 31, 33; and fireplaces, 17, 30, 31; ovens, 59
Canadian Pacific Railway, 81
canoes, birchbark, 49
cant hooks, 27, 28, 43

Cape Breton Island, 9
capstans, steam-powered, 28
cedar, 9
Champlain, Lake, 81
Chatham, New Brunswick, 52, 69, 81, 88
Chats Falls, 64
Chaudière Falls, 54, 55, 56, 57-58, 63, 64, 71
circular saws, 74, 74-75, 76
coal oil, 24
Cobalt, Ontario, 30
Cockburn, John, 49
company stores, 86
Connor, Ralph, 60, 66
conservation efforts, questions about, 89-90
contracts, 36-37
cooks and cooking, 17, 32, 34; in shanties, 31; on river drives, 49; on rafts, 59, 62
corruption, 86
Coteau Rapids, 62
Coulonge River, 43
cribs, on the Ottawa, 53, 54, 55, 56, 61; on St. Lawrence, 61
crosscut saw, two-man, 22, 23-24
crosscutting, 74
Crown lands, 15, 16, 17, 19, 65; care of, 92; and revenue, 86; timber poaching on, 86
Cunard, Joseph, 68, 85, 88
Cunard, Sir Samuel, 88
cutting rights, private, 16, 17

dams, use of for driving, 41-42
Dartmouth, Nova Scotia, 14
deacon seat, 31
desertion, 19, 36-37, 66.
Dodge, Arthur M., 87-88
double bobsleds, 26
double-edger saw, 73
Douglastown, New Brunswick, 30, 52
dram (of cribs), 61-62
draveurs. *See* river drivers
drive and boom companies, 63
drive camps, 46, 48, 49
drive streams, 41, 42, 43; clearing of, 41-42, 43
drivers. *See* river drivers
driving. *See* river driving
drownings, 45-46, 49, 52, 57-58

E.B. Eddy company, 71
economics of business, 42, 43, 47, 50, 54, 56, 57, 69, 74, 85-86
Eddy, E.B., 71
Edward VII, King, 55-56
Edwards, W.C., 82, 86
eighteenth century, 9-10

electricity, 71
Englehart, Ontario, 30
Erie, Lake, 60

farmers, 18, 25, 65, 90
fees, for booming companies, 63; for using slides, 54
fellers, 8, 23, 32
felling, 22-24, 25, 39
Finland, 19
Finnish workers, 19, 25
fires, 30, 85; and modern prevention, 91-92; risk of, 69, 71
fishermen-lumberers, 18, 65
Fitzpatrick, Rev. Alfred, 36
flangers, 49
floatability, 41, 43
floats (New Brunswick), 52
flumes, 43
foremen, 17, 32, 38; and river drive, 43-44, 45, 46, 47; as pilots, 56, 62; duties of, 32, 37; on rafts, 62. *See also* pilots
forest industry: and impact on economy, 15-17; benefits of, 90; criticism of, 89-90; image of, 89
forest management, 92
forest resources, 9
forklifts, 92
France, 9, 12, 15
Fraser Brothers, 85
fraud, 86
French Canadians, 18-19, 20, 21, 62, 66
Frontier College, 36
Fundy, Bay of, 65

gang saws, 73, 74, 75, 76
Garden Island (in Lake Ontario), 61, 62
Gaspé peninsula, 19
George V, King, 55-56
Georgian Bay, 19, 60, 61, 65, 87
Gilmour mill, Hull, 82
government control of logging rights, 15, 16-17, 19
government investigation into corruption (New Brunswick), 86
government legislation, 65; against dumping of refuse, 92; and government slides, 64; and sustained yields, 92; on child labour, 82
government, and the industry, 15, 16-17, 19, 39, 43, 54-55
Grant, George M., 56
Great Britain. *See* Britain and the British
Great Lakes, 65, 81. *See also individual lakes*
hardtack, 34
hauling logs and timbers

out of the bush, 25-29, 39; mechanized, 28-29
hauling roads, 29
hewing, 24; and waste, 29-30
horses, 17, 21, 25-26, 27, 28, 52, 76
Hudson's Bay Company, 66
Hull, 54, 58, 71
Huron, Lake, 60, 65
hydroelectric dams, 91

identification marks, 39, 40, 41
immigrants, 19; and cheap transport, 90
immigration, 7, 9, 67
Industrial Revolution, 7, 8, 9
Ireland, 85
Irish, 19-21
Iroquois Falls, Ontario, 91

J.R. Booth Lumber Company (Ottawa), 12, 84
James Maclaren and Company, 83-85
Joggins, Nova Scotia, 65
joints (New Brunswick), 52, 53, 61, 62

kerfs, 75, 76
key-logs, 44
Kingston, Ontario, 61

La Tuque, Quebec, 62
labour brokers, 19
labour relations, 37-38, 52
Lachine Rapids, 62
LaHave River, 9
Lake of the Woods, 14, 19
larrigans, 46
laths, 73
Lièvre River, 43
limit-holders, 15; and identification marks, 39, 40; cooperation among, 40-41
liner, 24
live rollers, 72, 73
living conditions, 38
locomotives, steam powered, 76
log booms, 50. *See also* booms
log rolling, 43
loggers, 9, 18, 65; and improved quality of life, 90; in literature, 66; reputations of, 92
logging, costs and revenues, 17; technical advances in, 24
logging camps, supplying, 21-22
logjams, 92; breaking up, 43-45
log-rolling, 39
Long Sault Rapids, 62
Lord's Day Act, 49

Lower Canada (Quebec), 12-13, 14
lumber camps (shanties), 17, 24-25; conditions in, 31-38; growth in size, 38
lumber industry: and impact on economy, 14; growth of 67-68; technical advances in, 68-69, 73-76. *See also* sawmill technology
lumber kings, 85-88; characters of, 85, 86; ethnic roots of, 85; influence of, 85
lumber mills, 18, 24; steam powered, 13; water powered, 13, 14. *See also* sawmills
lumber pilers, 82
lumber towns, 14
lumber yards, advances in, 76-78
lumber, 8; as major export, 14; decreased need for building, 90; domestic market for, 79; export of, 67, 68, 78-81; for building, 9; hand sawn, 13; loading of, 76, 80; shipping of, 78-81; stacking of, 76
lumberers. *See* shantymen
lumberjacks, 18
lumbermen (owners), 10, 14, 17, 18, 32, 41, 79-81; and relations with workers, 37-38; exploitation of employees, 36; foreign-based, 85; lumbers kings, political connections of, 86

machinists (in water-powered mills), 81-82
Maclaren brothers, 83-85
Maclaren, James, 85
Madawaska Improvement Company, 42
Madawaska Valley, 9
Maritimes, 14, 16, 17, 36, 43, 46, 50, 51-52, 65, 67, 81. *See also individual provinces*
Marquell, William, 82
masts, 9, 10-11, 22-23, 24; as major industry, 11, 12
Mattawa, River, 45
McLachlin Brothers, 78
McMullen, Thomas Gotobed, 85
meals, 32, 33, 34; on rafts, 59
Mersey River, 9
Methodists, 36
Michigan, 60, 65
mill towns, 86-87
millponds, 63, 69, 70, 72
Miramichi Bay, 52

Miramichi fire, 30
Miramichi River, 9, 30, 51, 85
missions, 36
Montferrand, Joe, 66
Montreal River, 44
Montreal, 18, 53, 62
Moose River, 45

New Brunswick, 9, 11, 12, 14, 21, 30, 42, 62, 66, 85, 86, 88; sawmills in interior of, 81
New England, 10; as lumber market, 79, 81
New France, New Brunswick, 81
New York City, 65
Newcastle, New Brunswick, 30, 52
Niagara Falls, 61
North West Company, 66
Nova Scotia, 9, 14, 31, 85

oak, 9, 12, 57
Ontario, 9, 14, 16, 17, 18, 31, 36, 37, 38, 42, 44, 45, 50, 55, 65, 81, 85
Ontario, Lake, 60
Ottawa, 12, 18, 19, 49, 54, 55, 60, 77-78
Ottawa River, 20, 22, 40, 49, 52, 53, 54, 60, 63, 66, 81, 85, 90, 92
Ottawa Valley, 12-13, 14, 19, 20, 29, 29, 36, 52
oxen, 17, 21, 25, 26, 27

Parrsboro, Nova Scotia, 14
payloads, 56, 57, 62
peaveys, 27, 43, 44
Pembroke, Ontario, 49
penstock, 69
Petawawa, Valley, 9, 43, 49
Peterborough County, Ontario, 27
Pigeon River, 43
pike-poles, 27, 57, 64-65
piling yards, 73, 76, 77, 78, 92
pilots (rafts), duties and importance, 56-57, 59
pine, 8, 9, 10; as major export, 12; for masts, 11
pit saws, 13
planks, 12
pointer boats, 46, 47, 48, 49
Poland, 19
pollution, 73, 90
Port Royal, 13
power sources, 9; developments in, 68-69, 71-72
Presbyterian Church, 36
Prescott, Ontario, 62
price fluctuations, 85-86
Price, E.J., 85
Prince Edward Island, 14, 85
privies, 17
professional lumber workers, 18-19
Protestant missionaries, 36
pulp and paper industry, 90
pulpwood, 91
Quebec City, 6, 13, 18, 50, 51, 52, 53, 56, 57, 60, 61, 62, 92
Quebec (province), 4, 9, 10, 14, 16, 17, 18, 36, 50, 66, 68, 79, 85

rafting timber, 40, 51; in Maritimes, 51-52; on the Ottawa, 52-60; on the St. Lawrence, 60-62
rafts, ocean-going, 65; as hallmark of square timber industry, 60; life on, 58-59; in New Brunswick, 52 53; methods of propelling, 57, 58-59, 62; on the Ottawa, 53, 54; on St. Lawrence, life on, 62; towed, 52; wind-powered, 52
raftsmen, 9, 18, 51-62, 65, 91, 92; on St. Lawrence, 61-62; reputation of, 66. See also river-drivers
railroads , 28, 29, 60, 61, 76, 79-81, 86
raking teeth, 24
rapids, 53, 57, 62, 91
reforestation, 92
religion, 36
Reversing Falls, Saint John, 52
Richelieu Canal, 81
Rideau Canal, 19
river drive, 39-66; end of, 91
river drivers, 9, 39-51, 65, 91; and logjams, 43-44; reputation of, 43, 45, 48, 66, 92; routines and living conditions, 48-51
river systems, as source of power, 9; for shipping and transportation, 9
river-drivers, 18. See also raftsmen
road-making, 27-28
Robertson Raft, 65
Robertson, Hugh, 65
Rockland, Ontario, 82
royal commissions, 82, 86
running the slides, 55-56
runoff, 43

safety measures, 45, 46, 90
St. Andrews, New Brunswick, 74
St. Anthony Lumber Co., 87
St. Croix River, 51-52
Saint John River, 9, 11, 51, 63; and pollution, 90
St. Lawrence River, 9, 11, 52, 56, 61, 81, 92
St. Lawrence, Gulf of, 52
St. Maurice River, 62
St. Stephen, New Brunswick, 83
saloons, 60
Sault Ste Marie, 25
saw-deck, 72
sawlogs, 12, 24-25, 27-28, 51, 67-68, 74; booming of, 62-63; rafted, 62; sorting of, 62, 64; towing of, 62
sawmill technology, 86; advances in, 9, 68-69,
71-72, 73-76
sawmills, 9, 39, 50, 62, 63, 64; accidents, in, 82; and child labour, 82; and impact on economy, 14; and labour relations, 83-85; and seasonal employment, 82-83; and skilled work in, 25; and strikes, 83-85; as beginnings of unions, 85, 86, 90; early, 68; electricity powered, 71-72; growth of, 67-68; mechanized, 13-14, 81-82; new sources of energy in, 90; steam powered, 69-70, 72, 73, 82; turbine driven, 69, 71; union activity in, 83; water powered, 68, 69, 70, 72, 73, 82; workforces in, 81-83; working conditions in, 82, 83
saws, development in, 68, 73-74, 75-76
sawyers, 74, 81
Scotland, 85
seasoning (air drying) of wood, 76-77
settlers, 13
Severn River, 9, 61
shanties (lumber camps), 16, 17, 18; cooking in, 31; description of, 31-32; fireplaces in, 31; heating, 31; hygiene in, 33; life in, 31-38; lighting in, 33, 36
shantymen, 9, 18, 19, 20, 30; diet of, 34; and work in the bush, 21-38; life of, 31-38; recreation of, 35-36; reputation of, 66; workday of, 32, 33, 35
sheer booms, 41
Shiners, 19-20
shipbuilders, 11-12
shipping of lumber, 78-81
shipping ports, 6, 9, 39, 51, 52
silviculture, 92
Simcoe, Ontario, 50
single stick slides, 42
skid roads, 12, 26, 27, 29, 32
skidders, mechanical, 89, 90
skidding logs, 26
sleds, horse-drawn, 21-22, 24, 25, 26, 27, 28, 29
slides, 43, 53-56, 62; government owned, 57
sluice gates, 56
Snowball, Jabez Bunting, 85
sorting booms, 62, 64
sorting timber and logs, 41, 65
South America, 79
Southwest River Log Driving Company, 42
splash dams, 42
spruce, 9
square timber, 9, 10, 11, 12, 22-24, 25, 26, 39; as major export, 12; cutting
and preparing, 24, 29-30; decline of industry, 67; rafting of, 52-60; shipping of, 51, 52; trade on St. Lawrence, 61
stamp hammers, 20
stamped markings, 20, 64; identification, 39, 41
steam logger, 28-29
steam power, 9
steamboats, 11, 21, 22, 57, 61, 62
Stiles, Johnny, 45
stockholders, 85
storing and shipping lumber, advances in, 76
St-Pierre, Lake, 53
strikes, 38, 83-85
stumpage fees (timber duties), 15-16; losses of, 86
supplying the bush camp, 21-22, 50
Sweden, 19

tamarack, 9
tea, 34-35
teamsters, 26, 32, 77
Temiskaming, Lake, 52, 53, 56, 63
theft, 41
Thomson, Tom, 49
Thunder Bay District, 38
tides, as power source, 14
timber barons, 85-88; ethnic roots of, 85; influence of, 85; political connections of, 86
timber berths (timber limits), 15
timber cruisers, 17
timber duties (stumpage fees), 15-16
timber limits (timber berths), 15, 17, 19, 86; bought by Americans, 65
Timber Marking Act, 41
timber rafts, 50
timber, loading, 6
timbering, in twentieth century, 13
timbermen, 20-21, 41, 56-57, 60-61. See also lumbermen
timing, 56-57
tolls, 43, 54
tools, 20, 23, 24, 27, 28
tote roads, 21
towboats, 21
tramcars, for moving lumber, 76
Trent River, 9, 54
Trent River Valley, 37, 60
truck pay (scrip), 86
trucking and trucks, 29, 91, 92
tugboats, 50, 52, 53, 63
Twain, Mark, 56
twentieth century, 18, 19, 38, 76, 81, 83, 88. 90

unions, 38, 52, 83-85
United States, 14, 67; as lumber market, 79, 81
Upper Canada (Ontario), 12-13, 14

Upper Ottawa Improvement Company (ICO), 63-65

van (wogan or wanigan), 36
violence, 16, 19-20, 36, 37, 42, 50, 83, 88
voyageurs, 65-66

wages, 32, 37, 39, 50, 52, 56, 57, 81, 83; loss of, 50-51
wagons, horse-drawn, 21
waney timbers, 30
wanigan (wogan or van), 36
War of 1812, 11
waste, 30, 73
waterfalls, 43, 53, 54, 57, 62, 63, 69; as power source, 9, 14
Waubaushene, Ontario, 87
wedges, 24
Welland Canal, 60-61
West and Peachey, 50
West Indies, 79
western provinces, 36
Weymouth, Nova Scotia, 14, 81
whipsaws, 13
Whitney, Ontario, 87
winches, 27, 49, 50
wind power (for sawmills), 14
windlasses, 44
Winnipeg, 81
wogan (wanigan or van), 36
wood, seasoned, 76-77
working conditions, 72
Wright, Philemon, 40
Wright, Ruggles, 53-54